I0161028

Logic's Dilemma

What is Truth?

First Edition

Logic's Dilemma: What is Truth?

© Copyright 2021 by The Heart of Man Publications, LLC

Copyright Office Registration Number: TXu-2-269-101

ISBN: 978-1-7368598-2-7

Until and unless notified, copies of the material found herein are authorized only by students, instructors, and others solely for personal use. To include any original material or original presentation of material contained herein in a publication requires prior written approval from The Heart of Man Publications, LLC.

All Rights Reserved. No part of this book may be reproduced, stored in a retrieval system, transmitted, or translated into machine language, in any form or by any means, electronic, mechanical, photocopying, recording, or otherwise, without the prior written permission of The Heart of Man Publications, LLC, except for brief quotations embodied in research articles and reviews. Any permissible public use must cite this book as a source.

Other books by The Heart of Man can be found at:

The Heart of Man Publishing – Your Source for Anonymous Publishing (www.roadtoatoe.net):

The Resurrection of Man: The Story of Adam and Eve Told Mathematically.

Table of Contents

Message to the Reader ix

The Beginnings x

Preface: The Role of Logic xii

Chapter 1: Philosophy of Logic 1

1.1 Logical Realism..1

1.2 Logical Nominalism..3

1.3 Nominalism vs. Realism in Logic 4

1.4 Difficulties in Logic 6

    1.4.1 Ideas vs. Sentences...6

    1.4.2 The Trouble with Universals................................7

1.5 Concluding Remarks 8

Chapter 2: Propositional Logic..10

2.1 Propositional Logic as the Study of Logical Forms 10

    2.1.1 Syntax and Semantics..11

2.2 The Object Language of Propositional Logic 12

    2.2.1 Logical Connectives and Formulas 12

    2.2.2 Logical Separators..13

2.3 Syntactical Rules of Propositional Logic......................13

    2.3.1 Tautology, Contingency, and Contradiction 13

    2.3.2 Truth Tables..14

2.4 Rules for the Five Fundamental Connectives 15

    2.4.1 Negation...15

    2.4.2 Conjunction..15

    2.4.3 Disjunction 15

    2.4.4 Implication 16

    2.4.5 Bi-Conditional 18

    2.4.6 Comments on Connectives.................................19

2.5 The Meta-Language of Propositional Logic 19

   2.5.1 Contextually Equivalent Propositions 19

2.6 Argument Forms 20

   2.6.1 Valid Argument Forms 20

2.7 Logical Inference 22

   2.7.1 Valid Inferences..23

      2.7.1.1 Example of a Sound Argument 24

      2.7.1.2 Conditional Proof 25

      2.7.1.3 Identifying Inconsistent Premises 26

      2.7.1.4 Reductio ad Absurdum 27

2.8 The Algebra of Propositions 28

2.9 Concluding Remarks 29

Chapter 3: Predicate Calculus 31

3.1 Terms and Predicates...32

   3.1.1 Open Sentences and Free Variables 33

3.2 Quantifiers..33

   3.2.1 Universes..34

3.3 Formulas in Predicate Calculus..34

   3.3.1 Making Sentences 35

      3.3.1.1 Valid Inferences within Predicate Logic 36

      3.3.1.2 Model Interpretation...36

3.4 Valid Argument Forms within Predicate Calculus 38

   3.4.1 Developing Valid Arguments 39

      3.4.1.1 Step 1: Symbolizing...39

      3.4.1.2 Steps 2–4: Dropping and Adding Quantifiers.................40

   3.4.2 Universal Specification and Generalization.............................40

   3.4.3 Existential Specification and Generalization 41

      3.4.3.1 Rules for Using Ambiguous Names 42

3.4.4 Additional Rules of Inference 42

3.4.5 Summary of the Rules of Inference 44

3.5 Predicate Calculus with Identity 45

3.5.1 Rules Involving Identities 45

3.5.2 The Predicate Calculus and Mathematics 46

3.6 Concluding Remarks 47

Chapter 4: Logic and Set Theory 48

4.1 Set and Logic Relationships 49

4.2 Well-Defined Sets 51

4.2.1 Set Specifications 51

4.2.2 Element and Set Associations 51

4.2.3 Complimentary Sets 52

4.2.4 Classes 52

4.2.5 Universal Set and the Null (Empty) Set 53

4.3 Set Operations 54

4.4 Comparing Set Theory to Logic 55

4.5 Concluding Remarks 58

Chapter 5: The Theory of Realist Logic 60

5.1 The Foundations of Logical Systems 61

5.2 The Syntax and Semantics of Realist Logic 61

5.2.2 Statements (The Semantics of Realist Logic) 62

5.2.3 Systems 62

5.2.3.1 Consistent and Inconsistent Systems 63

5.2.3.2 Complete and Incomplete Systems 64

5.2.3.3 Equal and Equivalent Systems 65

5.2.3.4 System of Systems 66

5.3 Complete and Consistent Systems 67

5.3.1 Valid Systems 67

5.3.2 Complete and Consistent Invalid Systems.............................68

5.3.3 Complete and Consistent Partially Valid Systems.................68

5.3.4 Proposition on Complete and Consistent Systems 68

5.4 Characteristics of a System 69

5.4.1 Subsystems...69

5.4.2 Universal Systems..70

5.4.3 Propositions on Systems 71

5.5 Set Operations on Valid Systems 72

5.6 Functions on Systems 74

5.6.1 Composite Functions on Systems................................75

5.6.2 Valid Functions 75

5.7 Relations 77

5.7.1 A Valid System of Relations..77

5.8 The Laws of Valid Systems 78

5.8.1 Valid Operators...80

5.9 Paradoxes 81

5.10 Concluding Remarks 83

Epilogue...85

Appendix..89

List of References..90

Message to the Reader

Realism is the philosophical position that universals are just as real as physical, measurable material. Nominalism is the philosophical position that universal or abstract concepts do *not* exist in the same way as physical, tangible material. Debating the merits of these two views greatly influenced multiple areas of thought throughout the Middle Ages and was crucial in developing theological scholarship [286].

To the logical realist, mathematical truths reside in the mind, independent of the brain, meaning they are discovered. Conversely, the logical nominalist argues that mathematics is a construct of the human brain. As an invention, mathematics only stipulates the use of certain symbols, much like how the rules of chess govern how the pieces can move. This book discusses realism vs. nominalism within systems of logic.

Before starting the first chapter, the reader is advised to review "The Beginnings" (pages x-xi). A brief description there explains how mathematical symbols are used throughout the book. Hopefully, familiarization with these symbols will lead to a more enjoyable reading experience.

— *The Heart of Man*

The Beginnings

Logic		
Description	**Symbol**	**Remark**
Proposition	Lowercase letters "(p, q)"	p: A declarative sentence that can be true or false.
Atomic proposition	Lowercase letters "(p, q)"	p: A fundamental proposition—a sentence or complete thought that cannot be made into a simpler thought.
Universal quantifier	$\forall$	$\forall x$: A quantifier that quantifies over every value of x.
Existential quantifier	$\exists$	$\exists x$: A quantifier that quantifies over at least one value of x.
Predicate variables	Uppercase letters, "W"	$\forall x(Wx)$: All $x's$ are $W's$.
Terms	Lowercase letters, "x"	$\exists x(Wx)$: There exists an x that is a W.
Factorial	$number!$	$n!$: $n(n-1)(n-2)\cdots 1$.
Infinite	∞	Infinity.

Set Theory		
Description	**Symbol**	**Remark**
An element of a "set"	Lowercase letters, "m"	m is a distinct element or member of a set.
Set	Uppercase letters, "$\{m\}$"	$S = \{m\}$: S is a collection of distinct elements, m, into a whole.
Subset of, not a subset of	$\subset, \not\subset$	$A \subset B$: Every element in set, A, is also in set, B. $A \not\subset B$: A is not a subset of B.
Set inclusion, exclusion	$\in, \notin$	$s \in S$: s is an element of S; $s \notin S$: s is not an element of S.
Empty or null set	$\emptyset$	$S = \emptyset$: S is the set with no elements in it.
Number of elements in a set (cardinality)	#	$\#S$: The number of elements in the set, S.
Relationship between the elements of two sets	$\xrightarrow{r}$, $\quad r(s) = \bar{s}$	$S \xrightarrow{r} \bar{S}$: The elements of S are related to the elements of $\bar{S}$ through a relationship, r.
Uniqueness	!	$!a$: a is a unique element.
Cross product of two sets	$\times$, "$A \times B$"	$\forall x \forall y (\langle x, y \rangle \in A \times B)$, $x \in A$, $y \in B$: All the ordered pairs, $\langle x, y \rangle$, where $x \in A$ is the first element, and $y \in B$ is the second.
Relations	"$R = (A, B, C)$"	If $x \in A$, $y \in B$ and $\langle x, y \rangle \in C$, then $C \subset A \times B$.
Functions	"$f(t)$"	$f(t)$: $\langle t, f \rangle$ is a relation such that for each t, f has one, and only one, value.

Preface: The Role of Logic

To discover truths is the task of all sciences; it falls to logic to discern the laws of truth. ... I assign to logic the task of discovering the laws of truth, not of assertion or thought.

– Gottlob Frege

When is it possible to definitively say, "this is a number?" Even conceding, philosophically, that an external world or reality exists independent of ourselves, numbers would not necessarily be part of that reality. So, from where do numbers come, and why do they play the specific roles in human life that they do?

There are essentially two perspectives on this question:

1. Numbers are inventions.
2. Numbers are discovered.

In either case, numbers come from someplace other than the sensorially accessible empirical world. Yet, ironically, scientists describe the nature of the empirical world almost exclusively by using numbers that lie outside of it.

Whether invented or discovered, numbers are abstractions, that is, objects that do not exist in the empirical world of space and time. Abstractions consist entirely of ideas, devoid of sensations, and thus are inaccessible to the senses. For example, the statement, "this is a tree," enjoys a certain level of concreteness as the object referred to as a tree exists in space and time. A "tree" has shape, size, color, and smell, all of which change with time, altogether accessible to the senses. Some philosophers argue that numbers are similarly concrete because they are "marks on paper," which indeed do exist in space and time. However, words (marks on paper) are not generally considered artifacts of space and time but rather abstractions. Words can represent empirical sensations, like colors (this is "red") or shapes (this is "fat"), but the words themselves are abstractions. So it is with numbers. Even idealized geometrical objects like circles, spheres, rectangles, and squares are, in a sense, abstractions.

If inventions, the linguistic philosopher inherits the task of explaining how humans write down specific configurations of words and numbers and then elicit meaning from them. As other species notably lack this ability, why do humans have the capacity to recognize a certain string of numbers and words as a meaningful sentence while deeming others as nonsensical?

If discovered, then where do words and numbers reside, and who put them there? It may be tempting to infer that humans construct a language by forming strings of words that correspond to certain states of affairs or facts experienced in the physical world. The strings of words acquire meaning by being matched to specific states of affairs. Yet this does not quite seem the case. Humans can string together intricate mathematical statements, considered meaningful, without connecting them to the physical world of experience. According to this view, numbers and word configurations exist in an independent world of "forms." Forms are entities meaningful to humans. Without "form," strings of numbers and words are merely meaningless marks on paper. Evidently, a human can recognize forms, such as "subject-verb-predicate," as meaningful while dismissing other word and number configurations as being not meaningful.

Practitioners who believe the rules of logic are invented are commonly called logical nominalists. Logical realists, meanwhile, are those who believe the rules of logic are discovered. To illustrate the stance of a logical nominalist, take the sentence, "There are three cows in the field." The meaning of this sentence seems evident enough for everyday conversation, but a nominalist might argue that such a statement is incomprehensible. Nominalism inclines towards believing that everything that exists does so in space and time. The sentence "there are three cows in the field" contains a number — an abstraction. Abstractions do not exist in space and time. There is nothing in the world of empirical experience, ordered spatially and temporally, that would suggest "number." There are not "three" of anything. There is simply a conglomeration of sense data viewed by an observer, yet nothing within it would suggest a number. A nominalist would likely not deny numbers but rather their ability to describe something about the physical world. According to the nominalist, employing abstractions in describing the physical world misrepresents what is truly experienced. Nothing that exists in space and time is a number. Ironically, scientists describe the physical world's nature almost entirely through numbers. According to the nominalist, numbers are inventions having nothing to do with a discoverable physical world.

The nominalist would likely further remark that, in everyday conversation, humans often use nonexistent universals when speaking or writing, which leads to the inappropriate use of numbers. For example, in the statement, "there are three cows in the field," the word, "cow," represents a universal. Hence, many cows can roam the fields, but there is only one "universal form" of a cow. However, to a nominalist, cows are particular objects, consisting of particular configurations of sense data, nothing more. Thus "cowness" as a universal form simply does not exist. Accordingly, a nominalist would argue that a realist who thinks a "cow" is something tangible is committing a philosophical blunder.

Here, the nominalist encounters practical difficulties. Universals are commonly employed in everyday speech. No one has yet developed a language devoid of words or symbols suggesting universals. People who use words like "dog," "cat," and "house" would claim to understand these words perfectly. Regardless, the nominalist would hold that such words are meaningless, misleading, refer to nonexistent worldly objects, and further obfuscate the nature of the physical world. At best, terms that suggest a universal are but habits of thought, a practice that assigns the same name to objects with similar appearances. While such an approach to language might be convenient, a nominalist would likely argue that such expediencies are only practical, not objective or factual.

Worse yet, mathematics is riddled with universals. The most significant is the "set," denoted as "$S = \{m\}$." The set is a collection of distinct elements, "m," into a whole, "S." While the elements of a set are usually numbers, they can be anything arising from thought or intuition. The philosophical justification for grouping distinct objects into a whole universally is not as clear as with ordinary language. Despite this uncertainty, virtually the entire body of mathematics is predicated upon this universal idea.

It is widely agreed that numbers emerge from logic, which begs the question, "where does logic come from?" In the Western world, the birth of formal logic is principally attributed to Aristotle. Based on the syllogism, his entire logical system is synonymous with deductive reasoning, wherein facts are determined by combining existing statements. This contrasts with inductive reasoning, wherein facts are determined by repeated observations [158].

After Aristotle, logic theory stagnated until Leibniz divided all true propositions — including mathematical ones — into two types:

1. Truths of fact.
2. Truths of reason.

For Leibniz, mathematical propositions fall into the second camp as their denial would be logically impossible. On the other hand, empirical propositions containing mathematical terms, such as "$2\ cats + 3\ cats = 5\ cats$" are truths of fact because they hold true in the actual world. According to Leibniz, this is because the actual world is the "best possible" of all worlds. Thus, while $2 + 3 = 5$ is true in all possible worlds, $2\ cats + 3\ cats = 5\ cats$ could be false in some world [159]. Leibniz was among the first thinkers to differentiate between developing logic systems and the discovery of worldly facts. His classifications suggested that logic systems were, in some sense, inventions.

With inspiration from Gottlob Frege (1848–1925) and Giuseppe Peano (1858–1932), modern logic emerged from the works of the aristocrat Bertrand Russell (1872–1970) and Alfred North Whitehead (1861–1947) in the *Principia Mathematica*, a three-volume text published in 1910, 1912, and 1913. The volumes endeavored to describe a set of axioms and inference rules in symbolic logic from which all mathematics originates. It was the foremost advocate that such an undertaking was attainable. Regrettably, the ambition collapsed when Kurt Gödel (1906–1978), the incomparable logician, proved that such an approach always produced statements not deducible from the logic of *Principia Mathematica* [160]. Nevertheless, most theorists regard mathematics and logic as identical branches of learning.

While Russell's chief concern was the rules of logical inference, others turned their attention to answering the question, "what is a number?" As geometry, arithmetic, algebra, and probability and statistics effectively address many practical problems outside the scope of logic, questions remained as to why these useful subjects should be called "mathematics." Descartes made some progress on the question by combining geometry and algebra into "analytic geometry." To make matters worse, Newton invented differential calculus that explained his "System of the World," which employed the concept of an infinitesimal as the basis for the ideas expressed in his book. At the time, most philosophers denied that infinity was in any way related to numbers. There was much debate over whether Newton's ideas were sound reasoning or a devilish deception leading to ruinous false beliefs. Lord Berkley was one of the fiercest believers that Newton's idea of "fluxions" (infinitesimals) was the ruinous kind [287].

In the 5th century BC, Zeno of Elea used the infinitesimal to prove that motion was impossible, which became known as the "no motion paradox." However, despite contrary empirical evidence, Zeno's contentions remained unrefuted until Georg Cantor's (March 3, 1845–January 6, 1918) 1874 paper, "On a Characteristic Property of All Real Algebraic Numbers" [161, 162]. Cantor was able to resolve the no motion paradox [288]. Cantor's paper, and his subsequent similar writings, eventually became known as "set theory." Set theory, in its axiomatic form, is widely regarded as the definitive work in the foundations of mathematics.

Cantor's main contribution was to interpret infinity numerically. The number of distinct elements in a set defines its "cardinality." For instance, the set, $H = \{1, \dots, 100\}$, has a cardinality of 100 since there are 100 elements in the set. All other sets containing 100 elements have the same cardinality as H. Determining the cardinality of an infinite set is more problematic. Consider the following set, $\mathbb{N}$, of counting numbers:

$$\mathbb{N} = \{1,2,3,\dots,n,n+1,\dots\}.$$

Since $\mathbb{N}$ has no largest number, it has an infinite number of distinct elements. How then can its cardinality be determined? Removing the even numbers from $\mathbb{N}$ leaves its cardinality unchanged; it has the same amount of numbers as before the even numbers were removed. Eliminating the first 10 integers, or even the first "M," where M is a large positive integer, would also leave its cardinality unchanged. Again, it has just as many numbers as before the M integers were removed. An infinite set is often defined as one equivalent to a proper subset of itself [179].

Cantor approached this problem by using a somewhat fabricated number to indicate the cardinality of $\mathbb{N}$, which he specified as "alpha null" ($\aleph_0$). He further proposed that, if the elements of an arbitrary set were in one-to-one correspondence with the elements of $\mathbb{N}$, that set would also have a cardinality of $\aleph_0$. The notion of one-to-one correspondence also held for finite sets. Two finite sets with the same number of elements have the same cardinality.

Many infinite sets have a cardinality of $\aleph_0$ — the set of integers, of rational numbers (fractions), and of algebraic numbers — a number that is a root of a non-zero polynomial in one variable with rational coefficients. Cantor then advanced his ingenious diagonal argument that proved, by contradiction, that the set of real numbers has a cardinality greater than $\aleph_0$ [164]. The set of real numbers, "$\mathbb{R}$," is more numerous than $\mathbb{N}$. Thus Cantor had shown, for the first time, that infinite sets revealed different cardinalities. The cardinality of $\mathbb{R}$, sometimes referred to as the "continuum," is designated "$\mathfrak{c}$" (a lowercase fraktur script "$\mathfrak{c}$").

Cantor also showed that, regardless of proximity, an infinite amount of real numbers exists between any two real numbers, $a < b$.

Furthermore, he demonstrated that these are just as numerous as those in the whole set of real numbers.

And yet, is there an infinite set whose cardinality, say "$\mathfrak{H}$," is such that $\aleph_0 < \mathfrak{H} < c$? Cantor hypothesized that no such set existed — known as the continuum hypothesis — although he could not conclusively answer this question during his lifetime. Ultimately, Gödel and Paul Cohen (April 2, 1934 — March 23, 2007) proved that the continuum hypothesis was unresolvable within set theory [165].

At the turn of the 20th century, Cantorian set theory was shown to contain several paradoxes (contradictions). Bertrand Russell and Ernst Zermelo (1871–1953) independently found one of these paradoxes (now named "Russell's Paradox"), which states that the set of all sets that are not members of themselves cannot exist. Such a set would simultaneously be and not be a member of itself. Cantor himself discovered "Cantor's Paradox" in 1896, stating that the "power set," the set of all subsets of a set, S, cannot be a member of S. However, consider the set, "U," of all sets. The power set of U would, by definition, be a member of U.

Paradoxes such as these arise because certain elements in set theory play a dual role. For example, some statements relate set elements to the properties that define a set and are also elements of the set. These conflicts confuse the object language, which discusses the individual elements, with the meta-language, which discusses the sets themselves. Russell's discovery led to further research into forming a clearer definition of sets, classes, and membership.

Russell addressed his paradox through the "theory of types," which clarified the distinction between the object and meta-language. All statements are placed into a hierarchy, or orders, according to their level of subject matter, thereby avoiding any reference to the set of all sets that are not members of themselves. Therefore, first-order logic quantifies over individuals, second-order over sets of individuals, third-order over sets of sets of individuals, and so on. To confuse orders is to commit a category mistake.

While Russell's theory of types circumvented all known paradoxes, it could not support the development of mathematics unaided. First, the infinity of natural numbers, or the idea that each natural number has a distinct successor, remains unproven within the theory of types. Second, the theorem that every bounded set of real numbers has a least upper bound, upon which the whole of mathematical analysis rests, is not derivable without further ad hoc strengthening of Russell's theory [159].

The paradoxes of Cantorian set theory had little impact on its success. Crucial contributions to the theory of the infinite far exceeded the inconvenience of dealing with the paradoxes it birthed. Moreover, the paradoxes of "naïve set theory" exist primarily on the subject's periphery, where practicing mathematicians rarely ply their trade. Unless research takes mathematicians to the subject's extremities, encountering a contradiction would be rare indeed.

While not disastrous, the paradoxes were embarrassing. Not long after their appearance, logicians contemplated ways of removing them. An analysis revealed that naïve set theory originated from three axioms:

1. **The axiom of extensionality**: Two sets with the same elements are identical.
2. **The axiom of abstraction**: Given any property, there exists a set whose members have that property.
3. **The axiom of choice**: Specify any collection of mutually disjoint non-empty sets, and a new set can be assembled containing exactly one element from each member of the given collection [167].

The axiomatic approach begins with a set of premises (statements accepted as true) and then proves as many theorems as possible from those axioms. Once progress ceases, additional axioms produce more theorems. The process continues like this until most mathematics develops fundamentally.

With refinements by Thoralf Skolem (May 23, 1887–March 23, 1963) and Abraham Fraenkel (February 17, 1891–October 15, 1965), Zermelo–Fraenkel set theory with the axiom of choice replaced naïve set theory as the definitive foundational work in mathematics. In essence, it is a unification of set theory with the predicate calculus and its extensions. It is generally agreed that Zermelo–Fraenkel set theory succeeded in developing the foundations of mathematics free of the paradoxes contained in naïve set theory. However, it requires approximately nine axioms (compared with naïve set theory's three) and excludes the concept of a "universal set" [256].

Whether dealing with integers, rational, real, or complex numbers, the following rule for multiplying two numbers is fundamental:

> Given ab and $a'b$ if $a \neq a'$, then $ab \neq a'b$ is true, which holds for every b (integer, rational, real, or complex number) except "zero." If $b = 0$, then $ab = a'b$, $a \neq a'$, which directly contradicts the rule guaranteeing uniqueness in multiplication.

Should zero be considered a number? There are two possibilities:

1. Declare that zero is not a number.
2. Declare that zero is a number with unique properties, where mathematical expressions involving zero, such as "$a/0$," are not numbers.

It is virtually universally accepted that zero is a number. Thus, the general approach has been to regard zero as a number and maintain that such expressions as "$a/0$", "$0/0$," or "0^0" are undefined.

Is there a relationship between zero and infinity? For instance:

$$\frac{1}{\frac{1}{100}} = 100 < \frac{1}{\frac{1}{1000}} = 1000 < \frac{1}{\frac{1}{10000}} = 10000 < \cdots$$

Since $1/1000$ is closer to 0 than $1/100$, and $1/10000$ is closer to 0 than $1/1000$, it might be reasonable to conclude that:

$$\lim_{a \to 0} \left(\frac{1}{a}\right) = \infty.$$

Somewhat problematically, this immediately raises other, more complex, questions. For example, the equations, $1/R = 0$ and $R + 1/R = 0$, are genuinely troublesome. Attempting to solve $1/R = 0$ algebraically, assuming $R \neq 0$, gives the unsupportable result, $1 = 0$:

$$\frac{1}{R} \cdot R = 0 \cdot R \to 1 = 0.$$

To assume that $R = 0$ leads to $1/0 = 0$, which contradicts the contention that $1/R$ approaches infinity as R approaches zero. Moreover, this also leads to the unsupportable:

$$\frac{1}{0} \cdot 0 = 0 \cdot 0 \to 1 = 0.$$

Concerning the second equation, $R + 1/R = 0$, if $R = 0$, then:

$$0 + \frac{1}{0} = 0 \to 0 \cdot 0 + \frac{1}{0} \cdot 0 = 0 \cdot 0 \to 1 = 0.$$

Again, this is an unsupportable result. Likewise, as $1/R$ gets smaller, R grows uncontrollably, giving $\infty \approx 0$. Solving the equation algebraically results in $R = \pm\sqrt{-1}$, which is not a real number.

The problem of nothingness arises in the very foundations of mathematics. In set theory notation, nothingness is signified by "Ø," the "null or empty set," defined as a set containing no elements. The problem involves the question of existence vs. property. Does the null set exist? Or is emptiness merely a property of a set? Naïve set theory often portrays the empty set as both. For example, the set of pink elephants is empty. In this context, the null set represents "nothingness" or "nonexistence" as a property. Elephants exist, but there are none with the property of "being pink."

On the other hand, the Loch Ness monster does not exist. Assuming this statement is true, the set of Loch Ness monsters is empty. In this case, Ø signifies the nonexistence of a substantial object, not the nonexistence of a property of that object.

The dilemma is not unlike the philosophical debate regarding substance vs. attribute. The substance exists, while the attribute is a substance's property, which can only exist because the substance does. At one stage, philosophers believed that there must be a substance to which properties or attributes belonged. However, the difficulty in describing substance led to philosophical systems that purged it entirely as a concept from the discussion. Whether or not this banishment will be maintained is a matter of debate [289].

The logicians who formulated axiomatic set theory recognized this conflict. The substance of axiomatic set theory is the set. Every object that exists in axiomatic set theory is a set. The null set exists and is unique. The idea of a property requires an additional axiom, the axiom of separation, which states that if p is a property and A is a set, the formula, $\{x \in A \rightarrow p(x)\}$, says that all the elements of A have property, p. A set of properties is a subset of the previously given set, A, which rules out Russell's paradox. However, the concept of a set becomes a universal in set theory, the very existence of which some philosophers deny.

In axiomatic set theory notation, a set is defined as:

$$y \text{ is a set } \leftrightarrow \exists x(x \in y \vee y = \emptyset).$$

A set is an entity that contains members or elements or is empty. A standard definition of the null or empty set is:

$$y = \emptyset \leftrightarrow \forall x(x \notin y).$$

This definition guarantees the existence of at least one set—the empty set. Based on this definition, any individual is identical to the empty set, which effectively excludes individuals [167]. Therefore, every object in axiomatic set theory is a set.

Producing whole numbers requires the axiom of infinity or axiom of pair, which provides other empty sets. This axiom reads:

$$\exists s \forall x \big(x \in s \rightarrow \{x, \{x\}\} \in s\big).$$

In other words, there is a set, "s," such that, if x belongs to s, so does $\{x, \{x\}\}$. Simply put, the set, s, has the property of reproduction. Sets that satisfy this property are inductive (infinite) sets. Whole numbers are constructed as follows:

$$\emptyset = \{\,\},$$
$$1 = \{\emptyset, \{\emptyset\}\} = \{\{\,\}, \{\{\,\}\}\},$$
$$2 = \{1, \{1\}\} = \{\{\{\,\}, \{\{\,\}\}\}, \{\{\{\,\}, \{\{\,\}\}\}\}\}, \ldots$$

In combination with the axiom of infinity, the existence of the null set guarantees these sets' existence [174].

The above is not without issue. Taking the definition of a set and inserting into it the definition of the empty set leaves:

$$y \text{ is a set } \leftrightarrow \exists x (x \in y) \vee \forall x (x \notin y).$$

However, the quantifier, "$\forall$," implies that the "$x's$" (every set) should include the empty set since it is a set. Moreover, the empty set is not a member of itself. Accordingly, the definition above must be modified to read:

$$y \text{ is a set } \leftrightarrow \exists x (x \in y) \vee \forall x (x \notin y \wedge x \neq \emptyset).$$

Nonetheless, such a definition would prove to be recursively self-referential. In other words, the empty set must already exist before its definition. To see this intuitively, suppose $x = $ "trunk" and $y = $ "tree," then the first part of the definition says that there exists a trunk that belongs to a tree. But, the second part of the definition declares that "or, there exists a tree to which no trunks belong." Such a statement begs the question, "what is the definition of a tree, then?" Presumably, having a trunk is part of what it means to be a tree.

To overcome this difficulty, in most axiomatic set theory systems, the existence of the empty set is given as an axiom.

Russell and Zermelo recognized that the paradoxes within Cantorian set theory arose within the axiom of abstraction, which reads:

$$\exists y \forall x (x \in y \leftrightarrow p(x)),$$

where y is not a free variable in $p(x)$ and which says that every x belonging to the set, y, has the property, p. To obtain Russell's paradox, let p be the property that x is not a member of itself, i.e., $x \notin x$. Allowing $p(x) = x \notin x$, then the axiom of abstraction reads:

$$\exists y \forall x (x \in y \leftrightarrow x \notin x).$$

However, a set of all sets must be a member of itself. Hence, in the formula above, put $y = x$, and the axiom of abstraction becomes:

$$\exists x \forall x (x \in x \leftrightarrow x \notin x).$$

By the rules of sentential logic:

$$\exists x \forall x (x \in x \leftrightarrow x \notin x) \equiv false.$$

A set cannot both be and not be a member of itself — Russell's paradox.

Overcoming the paradox requires substituting the axiom of separation for the axiom of abstraction, which reads:

$$\exists y \forall x \big(x \in y \leftrightarrow x \in z \wedge p(x)\big).$$

In the above, z is a set that is not a member of itself. Letting $p(z) = z \notin z$, then:

$$\exists z \forall z (z \in z \leftrightarrow z \in z \wedge z \notin z), \qquad z = x, y.$$

Note here that $z \in z \equiv false$ and $z \in z \wedge z \notin z \equiv false$. Therefore, by the rules of sentential logic:

$$\exists z \forall z (z \in z \leftrightarrow z \in z \wedge z \notin z) \equiv true.$$

However, the axiom of separation precludes the existence of a universal set. To see this, consider the definition of a universal set, U:

$$\exists! U \forall x (x \in U \leftrightarrow x \notin A).$$

This states that there is a unique set, U, that all other sets belong to, and no set belongs to any other set, A. The trouble begins with the definition of what is not in a set, i.e.,

$$\forall x (x \in {\sim}A \leftrightarrow x \notin A).$$

This says that a set exists, "${\sim}A$," to which all the elements, x, belong, and none of them belong to A. Hence, the formula above is reduced to:

$$\exists! U \forall x (x \in U),$$

since ${\sim}A = U$. However, since U is a set, one of the x's must be U. Hence $U \in U$, which is the source of Russell's paradox. There can be no universal set such that it is a member of itself:

$$\sim\exists U \forall x (x \in U).$$

The sentence, "$\emptyset = \{\,\}$," says that the null set exists. However, "$\emptyset = \{x : x = x\}$" suggests that some sets have the property of nonexistence, in this case, the universal set. In this context, $\emptyset$ indicates a set property rather than set existence, which follows because of $\emptyset$'s recursively self-referential definition. Axiomatic set theory can be formulated so that all sets take the form of "$\{x : p(x)\}$," where a set is defined:

$$\{x : p(x)\} = \left[y \leftrightarrow \exists x (x \in y \leftrightarrow p(x))\right] \vee y$$
$$= \emptyset \wedge \left[\sim\exists y \forall x (x \in y \leftrightarrow p(x))\right], \qquad p(x) = \text{"}y \text{ is a set."}$$

Therefore, axiomatic set theory declares that, unless the set is empty, existence is a property of a set. A non-empty set has the property of existence. It is hard to imagine that defining existence and nonexistence in this manner would not lead to the same difficulties in the foundations of mathematics as it does in arithmetic and philosophy. Intuitively, if the null set exists, there can be no universal sets. To reach this conclusion, suppose there is a non-empty set, "$C = \{c_1, c_2, ...\}$," where all the "$c_i's$" are sets of cows. If so, the only objects in the universe, C, are sets of cows. However, if the null set exists, at least one set is not a cow set, namely the null or empty set. Hence, $\emptyset \notin C$. Therefore, C cannot be a universal set because there exists a set that does not belong to it.

To the logical realist, mathematical truths reside in the mind, independent of the brain. Hence, mathematical truths are discovered. Conversely, the logical nominalist argues that mathematics is a construct of the brain, meaning that mathematical objects are invented. Thus, as an invention, mathematics only stipulates how certain symbols are used in a language, much like chess rules govern how its pieces move.

The celebrated mathematician David Hilbert (January 23, 1862– February 14, 1943) greatly influenced nominalistic formalism. In 1920, Hilbert proposed "Hilbert's Program," a research project which sought to formalize all mathematics by showing:

1. All mathematics follows from a finite set of axioms.
2. The axioms can be proven consistent.

The axioms provided the rules, and, if consistent, all mathematical operations could theoretically be derived and performed by a computer.

However, in 1931, Gödel presented the greatest challenge to the nominalist contention that mathematics is invented. He published two significant theorems in a paper entitled "On Formally Undecidable Propositions of 'Principia Mathematica' and Related Systems." He proved that for any computable axiomatic system powerful enough to describe arithmetic (e.g., the Peano axioms, Zermelo–Fraenkel set theory with the axiom of choice):

1. Any such consistent nontrivial logical system was incomplete.
2. Proving the axioms of the system consistent within the system was impossible.

Gödel's theorems provided evidence that mathematics cannot have been entirely invented, as "proof" and "consistency" rely on factors beyond the axioms of any nontrivial mathematical system. However, a nominalist would likely interpret Gödel's results differently. A nominalist may well accept that Gödel revealed mathematical systems' limitations but deny claims beyond that.

The philosophical pendulum embodied in the question "what are logic and mathematics?" has swung far to the side of the formalist/nominalist perspective—where mathematics evolves from creating formal rules unrelated to a discoverable world of mathematical truth. However, this begs the question, would a logical system formed from realist ideas—that mathematical truth resides in an independent mind—produce a significantly different logic/mathematical system from today's primarily nominalistic systems?

This book discusses both nominalist and realist approaches to building logic systems. Chapter 1 discusses the philosophy of logic and the turmoil it creates between logicians.

Chapter 2 then examines propositional logic, i.e., the study of deductive logical forms. Propositional logic is a complete logic system in which all possible conclusions follow from approximately 10 laws. However, such an apparatus is insufficient to express logical distinctions of the most prominent and elementary kind. The rules of arithmetic fail to materialize within propositional logic. Moreover, there is no means of symbolizing common grammatical distinctions, such as proper nouns, pronouns, adjectives, or adverbs. A richer set of notions is thus needed to support any group of systematic facts [178]. Such a system is first-order predicate calculus, sometimes called "mathematical logic," comprehensively explored in Chapter 3.

Chapter 4 examines the correlation between propositional logic, predicate calculus, and set theory. Axiomatic set theory, derived from a single set of rules, avoids all known paradoxes but eliminates a universal set. Would it then be possible to construct a logical system that avoids all paradoxes but includes a universal? While the answer may evidently appear to be "yes," this construction would require developing a logical system markedly different from nominalistic approaches. Such a logical system, developed in Chapter 5, adopts a realist approach to logic.

Chapter 1: Philosophy of Logic

God exists since mathematics is consistent, and the Devil exists since we cannot prove it.

– André Weil

Formal logic, which emerged in the Western world approximately 2,500 years ago, should be a well-substantiated field of study with only a handful of minor theoretical issues left unresolved. However, the subject's scope has expanded so dramatically as to include virtually all of mathematics. Consequently, many philosophical issues hindering the subject's growth have led to turmoil amongst logicians. Addressing these various issues, then, will help to uncover the true nature and utility of logic.

In this vein, the following sections will frame the question, "what is logic?" in terms of the philosophical perspectives of nominalist and realist logicians. The type of realism discussed is "Platonism," the theory that numbers and other abstract objects are objective, timeless entities, independent of the physical world and the symbols used to represent them. Additional perspectives on realism and nominalism not here discussed can be found in alternative sources giving contrasting viewpoints [291].

1.1 Logical Realism

Logical realism is the belief that logic is independent of language, thought, and practices. Generally speaking, a realist believes that mathematical objects exist regardless of whether they are thought about. For example, to a realist, if electrons and planets exist independently of an observer, so do numbers and sets, which means that ideas about electrons and planets are correct so long as they accurately describe the actual properties of real electrons and planets. A realist makes the same claim about numbers and sets. Thus, the logical realist believes that mathematical truths are discovered as opposed to invented [251].

1

Philosophers have developed various objections to logical realism, claiming that abstract mathematical objects are impenetrable and problematic. Consequently, logical realism has been among the most hotly debated topics in mathematical philosophy over the past few decades [251].

Logical realism can be characterized by:

1. **Existence**: Mathematical objects exist.
2. **Abstraction**: Mathematical objects transcend space and time.
3. **Independence**: Mathematical objects are independent of thought.
4. **Universal**: A shared property that transcends sense experience.

Existence is the idea that mathematical objects exist independently of language, thought, and practices, formalized as:

$$\exists x (x \in M = \text{"}x \text{ is a mathematical object"}).$$

The above is true regardless of whether anyone thinks about the mathematical object or not.

Abstraction is the idea that there are objects that exist outside the empirical world of space and time. Hence, mathematical objects can only be known, not experienced through the senses. Abstractions are often called "forms," which exist in an independent world of Platonic forms. An example of an abstraction is a number, which does not exist in space and time.

Independence is the idea that, regardless of intelligent beings and their language, thought, and practices, mathematical objects would still exist in their given form [251]. Hence, mathematical objects are timeless.

Finally, a "universal" is a common property possessed by a collection of objects that transcends sense experience. Logical realism affirms universals. A non-mathematical example of a universal is "cowness." There are many cows, but only one universal form of a cow. Such a form transcends the sense data associated with a particular object (cow).

1.2 Logical Nominalism

Nominalism comes in at least two varieties. The first rejects abstraction, while the second rejects the idea of a universal. The two varieties are independent, and either belief is consistently held without the other. However, the motivations and arguments underlying both are somewhat similar [252], which means that, hypothetically, a pure nominalist would deny both universals and abstractions.

Nominalism is not merely the rejection of universals and abstractions, though. If that were so, a nihilist, who rejects all entities, would classify as a nominalist. To be more specific, nominalism implies that everything is particular or concrete rather than trivial [252]. Thus, one kind of nominalism grants the existence of particulars, while the other asserts that everything is concrete. Put simply, everything exists in space and time.

Nominalism, in both senses, is a form of anti-realism. Notably, rejecting properties, propositions, possible worlds, and numbers as real, as well as any other non-spatial and non-temporal entities, would not automatically classify as a nominalist. Indeed, a nominalist must reject them on account of being universals or abstractions [252]. Unlike nihilists, both nominalists and realists agree that objects exist but disagree on their nature. A nominalist would not deny that cows exist, only that anything resembling a cow's universal form exists. A cow consists of a particular collection of sense data identifying them as a specific entity.

The nominalism discussed here rejects both universals and abstractions. However, the point is not to debate the merits of nominalism vs. realism but to compare and contrast two diametrically opposing views. In so doing, uncover the salient points with which to understand what logic truly is.

1.3 Nominalism vs. Realism in Logic

The debate between nominalists and realists about logic primarily revolves around semantics (the meaning of sentences). However, despite their disagreements, virtually all logicians agree that certain language forms (syntaxes) are part of logic. For instance:

1. All $S's$ are $M's$, all $M's$ are $P's$, therefore all $S's$ are $P's$.
2. $x = x$.
3. $p \wedge \sim p$.
4. $p \vee \sim p$.

However, there is considerable disagreement regarding the interpretation of these forms. For example, logicians who regard sets, classes, and numbers as being fabricated (invented) are called logical nominalists. Conversely, those who consider sets, classes, and numbers as independent objects are called logical realists. To illustrate the difference, a nominalist is not likely to say:

A. For all classes, "S, M, P," all $S's$ are $M's$, all $M's$ are $P's$, then all $S's$ are $P's$.

A nominalist would more likely say:

B. The following turns into a valid form no matter what words or phrases of the appropriate kind are substituted for the letters, "S, M, P," in the sentence, "all $S's$ are $M's$, all $M's$ are $P's$, then all $S's$ are $P's$."

The nominalist disputes the existence of such abstractions as classes. On the contrary, sentences, words, and letters are concrete, existing in space and time. Hence, the preference for the formulation, B, over A [157].

As to the form, "$x = x$," some philosophers contend that the symbol, "$=$," representing the reflexive relation, is not a relation, as relations make sense only if one object bears a relationship to a different object. Thus, they argue that whatever "$=$" represents, it is not a relation [157].

As to forms 3 and 4, the dispute concerns what p represents. Nominalists would likely claim that p is an appropriate sentence. At the same time, realists would find it patently absurd that logic relates to sentences at all, arguing instead that it concerns only with what the sentences say. Thus, to a realist, the assignment of a true/false value to p transcends logic.

Indeed, a realist would say, "the moon is spherical," if and only if the moon is spherical. The sentence, "the moon is spherical," represents an insubstantial idea. A true or false idea must be instantiated by an external (to logic) state of affairs, or facts. The instantiation, not the sentence itself, is what makes the sentence true or false. A sentence consists of comprehensible mounds of ink on paper, the disappearance of which would not erase the idea the sentence expresses. Thus, a realist would reject the nominalist contention that a sentence or a list of sentences alone is meaningful [157]. Moreover, the realist would likely reject the idea that sentences alone (marks on paper) are concrete.

The nominalist is apt to retort that one of logic's objectives is to find valid language forms independent of the meaning of the sentences that comprise the language. Such a structure is a "tautology." For instance, "$p \lor \sim p$" is a tautology, a language form always valid so long as p is an appropriate sentence, independent of what it says.

Such an argument would not satisfy the realist, though. A logical argument form, or schema, is a list of sentences, all of which are premises except one, the conclusion. In logic, there are valid and invalid argument forms. For instance, "all $S's$ are $M's$, all $M's$ are $P's$, therefore all $S's$ are $P's$" is a valid argument form, whereas, "some $S's$ are $M's$, therefore all $S's$ are $M's$" is invalid. Defining a valid argument requires the notion of "truth," a concept the nominalist would prefer to discard since it does not represent a concrete experience. A valid schema, S, is one where if the premises in S are true, so is the conclusion in S. Invalid schemas violate this rule. While the nominalist would like to purge logic of the abstraction, "truth," all attempts to write a language without abstractions have failed thus far.

1.4 Difficulties in Logic

Except for using different symbols, the rules associated with predicate calculus are practically identical to Cantorian set theory. Cantor defines a set as: "A collection of objects into a whole of our intuition or our thought. Hence, $M = \{m\}$ refers to a set, 'M,' comprised of elements, 'm.'"

Cantor's definition raises two questions:

1. Of what are the elements, "m," thoughts or ideas?
2. In what sense are the elements, "m," a collection?

1.4.1 Ideas vs. Sentences

To address the first question, the nominalist prefers that m refer to concrete objects perceptible through sense experience existing in space and time. If the elements, m, signify ideas, what do they consist of? From the nominalist perspective, ideas are thus identical to mounds of ink on paper. The nominalist argues that the brain can identify specific mounds of ink as a sentence while discounting a seemingly similar arrangement of words and letters as not being a sentence. However, an explanation of the mental capacity that facilitates this ability remains elusive. Moreover, such a declaration admits that the brain can distinguish between configurations of words that do and do not comprise a sentence, an abstracting capability that the nominalist generally denies.

The realist would further retort that ideas are not concrete objects. Ideas do not exist in space and time, and the fact that they are represented by mounds of ink on paper is irrelevant. Those mounds represent abstractions, not concrete objects. The mounds are meaningful only if correlated to something else. Therefore, it is not the sentences that are meaningful but what they express.

Further still, if the mounds correspond to an external state of affairs, the state of affairs would still exist independently of the mounds. Evidently, to the realist, this is what makes ideas true or false. Therefore, the string of words is true or false only if meaningfully correlated to an external state of affairs. Persuasively, words represent the idea of an external state of affairs, embodied in the relation:

$$\langle string\ of\ words, state\ of\ affairs \rangle.$$

1.4.2 The Trouble with Universals

Plato believed that forms were more real than ordinary physical objects. Unlike physical objects, they neither perish nor change and are not perceived through the senses but simply known. He also attributed certainty to them.

Moreover, Plato thought that forms corresponded not just to numbers but to general concepts, such as "tree," "horse," "yellow," "round," and "cold." Thus, individual instances of yellow are perceived, but the "form of yellow" or the "form of an individual horse" can only be known. Such general concepts are called "universals."

In what sense is $\{m\}$ a collection? Here, nominalist arguments stand on more solid ground than their realist counterparts. If $\{m\}$ signifies a collection of m numbers, then what does this mean? If the same name is assigned to similar objects, the nominalist would likely not object. However, the realist would argue that there is something called the "universal form" of a number, which implies that while there are many numbers, there is only one "number form." Here the nominalist would retort that nothing exists in space and time that would suggest a universal form.

If so, then where do universal forms reside? The realist would likely point to a world of forms similar to how physical objects are discovered through sense perception. Every class of objects has a universal form. For example, "catness," "dogness," "greenness," "humanness," and "numberness" are all universal forms.

A nominalist would undoubtedly reject these assertions. Objects that give meaningful substance to the world emanate from sense experience — objects in space and time, yet nothing in sense experience suggests anything approaching a universal form. In fact, a nominalist would say that the idea of a universal form runs counter to everyday experience and should have no place in a conversation about the real world.

Moreover, upon what grounds can a realist say, "there are m cows in the field"? From where does the idea of a number come? There is simply nothing in the concrete world of experience that would suggest the existence of numbers. Again, numbers are concrete only to the extent of being marks on paper. They are, therefore, inventions. There is nothing external to the human brain that a realist could point to and conclusively say, "that is a number." Such an object does not exist. See [157] for a detailed discussion of this issue.

1.5 Concluding Remarks

Classes, numbers, and sets will undoubtedly continue as components of logic. At some point, the nominalist must accept that the brain can abstract meaningfully to a certain degree. Still, such power is confined to thoughts engendered in the brain and does not extend to an external mind. According to this view, logic and mathematics represent the brain's ability to create, define, and recognize certain language forms. Valid and invalid logical forms are but types of language forms. To the nominalist, logic comprises formal linguistic rules that do not extend to an external world. However, nominalist arguments regarding how these linguistic forms manifest themselves as valid or invalid language forms are relatively weak.

The realist, conversely, demands the existence and independence of logical forms. Logical forms exist independently of ideas. Thus, a realist is likely to say:

> *a, b, and c and so on exist, and the fact that they exist and have forms such as F-ness, G-ness, and H-ness is independent of beliefs, linguistic practices, conceptual schemes, and so on* [250].

However, modern logic lacks a logical form independent of a linguistic form that a truly realist conception of logic would demand. The argument, "all S's are M's, all M's are P's, then all S's are P's," while valid, is simply a linguistic form. Even if the S's, M's, and P's have independent states of affairs instantiating them, modern logic lacks a separate language form representing this.

Logical realism asserts that logical entities exist in an independent human mind. Thus, mathematics is discovered, not invented. This point of view suggests a sort of mathematics where discoverable objects are represented independently of inventions.

Many working mathematicians have been, or are, logical realists; Gödel, for instance, believed in an objective mathematical reality perceived in a way analogous to sense perception. While Gödel was arguably the most celebrated logician of the 20th century, like Einstein, he sat on the sidelines, toiling in abstention. At the same time, the subject of logic precipitated the nominalist interpretations it currently enjoys. Gödel fought this trend by highlighting modern logic's severe limitations. He explained that if a consistent logic system could support arithmetic, it was incomplete. There would always be true but unprovable statements within any such system. Moreover, the axioms of the system could not be proven consistent.

Despite this, Gödel did not, in his lifetime, produce a truly realist system of logic. This book, then, seeks to answer the question, "what does a true realist system of logic entail?"

Chapter 2: Propositional Logic

Thus, mathematics may be defined as the subject in which we never know what we are talking about, nor whether what we are saying is true.

— Bertrand Russell, **Mysticism and Logic**

An argument is a series of statements ending in a conclusion. The premises which precede the conclusion either prove or provide evidence for its being true or likely true [175]. Arguments arise daily intended to promote one viewpoint over another. But which arguments should be accepted and which ones rejected? Being the science of valid arguments, logic provides the laws of truth and a guide for answering this question.

2.1 Propositional Logic as the Study of Logical Forms

Propositional logic is the study of deductive logical forms. For example, the following arguments have the same form:

> Today is either Monday or Tuesday.
> Today is not Monday.
> Therefore, today is Tuesday.

> Either Rembrandt painted the *Mona Lisa,* or Michelangelo did.
> Rembrandt did not paint it.
> Therefore, Michelangelo did.

> Either he is at least 18, or he is a child.
> He is not at least 18.
> Therefore, he is a child.

Such arguments take the form:

> p or q.
> Not p.
> Therefore, q,

where p and q represent declarative sentences that can be assigned a truth value (either true or false). Such sentences are called "propositions."

2.1.1 Syntax and Semantics

Systems of logic should have well-defined syntaxes and semantics. Syntax defines the structural rules of the language, and semantics defines what the rules mean [176]. In propositional logic, the object language provides syntax, stipulating its formal structure within the system. Semantics is associated with the meta-language, which discusses a sentence's meaning within the object language.

All sentences within the object language must be declarative, that is, affirmed as true or false. Statements like "what time is it?" or "close the door!" are not declarative since they cannot affirm truth or falsehood. Such statements operate outside the object language of propositional logic.

Propositional Logic	
Object Language	**Meta-Language**
$P = \{p_1, p_2, ..., p_n\}$	$R = \{=, \equiv, \in\}$
$C = \{\wedge, \vee, \sim\}$	$T = \{t, f\}$
$S = \{(\,), [\,], \{\}, ...\}$	

Table 2.1.1-1

Generally, logic is classified as either informal or formal. Roughly speaking, the former emphasizes semantics, while the latter concerns itself principally with syntax. Propositional logic is primarily formal. Table 2.1.1-1 summarizes propositional logic as a collection of sets. Each set, "P, C, S, R, T," associates with a unique property within propositional logic, explained in the following paragraphs[1].

---

[1] There may be some confusion in the notation here. The symbol, "=," is part of the object language of set theory, but is considered part of the meta-language of propositional logic. Care must be taken not to confuse the two uses.

2.2 The Object Language of Propositional Logic

The set, P (Table 2.1.1-1), represents part of the object language of propositional logic. The p_i's represent declarative sentences termed "atomic propositions." Generally speaking, atomic propositions are fundamental to more complicated sentences. All atomic propositions are declarative and irreducible, that is, incapable of being broken into simpler components. Put differently, reducing any atomic proposition into more fundamental pieces would result in a non-proposition. An atomic proposition is a sentence without connectives (discussed below).

Atomic propositions exhibit some critical characteristics. First, there is a functional relationship between the elements of P and T (Table 2.1.1-1). If $p \in P$, then $p(v) \in T$, where $p(v)$ signifies the semantics of p. The statement $p(v) \in T$ says that every atomic proposition has a truth value, $p(v)$. An atomic proposition must be either true or false by definition; what it says contextually is logically irrelevant. However, each proposition must make sense contextually.

Secondly, all atomic propositions are independent. Changing the truth value of an atomic proposition does not alter the truth value of a different atomic proposition. Notably, statements like $p \in P$ and $p(v) \in T$ are part of the meta-language. Such statements provide information about the object language and are therefore not propositions.

2.2.1 Logical Connectives and Formulas

A formula is a sentence that belongs to the object language. All propositions are formulas. The elements of C (Table 2.1.1-1), called "logical connectives," create additional formulas or modify existing ones. Logical connectives make compound propositions out of atomic propositions. Logical connectives are often referred to as "logical constants." The elements of C do not vary, unlike propositions, which are considered variable.

The following list shows examples of formulas in which p and q are atomic propositions:

$$p, q, p \vee q, \sim(p \rightarrow (p \wedge q)).$$

2.2.2 Logical Separators

The elements of S (Table 2.1.1-1) are called "logical separators." The compound propositions, $\sim p \vee q$ and $\sim(p \vee q)$, are formulas, though not identical ones. The separator, "( )," in the second formula modifies the first. The modified formula is associated with a different set of truth values [176].

2.3 Syntactical Rules of Propositional Logic

The syntactical rules for propositional logic are as follows:

1. Atomic propositions are formulas.
2. If ϕ and ψ are formulas, then "$\sim\phi$, $\phi \wedge \psi$, $\phi \vee \psi$, $\phi \rightarrow \psi$" and "$\phi \leftrightarrow \psi$" are formulas.
3. Formulas containing the proper use of logical separators are formulas.

Formulas constitute the object language of propositional logic. Thus, the meta-language is comprised of statements about formulas.

2.3.1 Tautology, Contingency, and Contradiction

There are three types of formulas:

1. Tautologies: Formulas whose truth values are all true.
2. Contingencies: Formulas whose truth values can be either true or false.
3. Contradictions: Formulas whose truth values are all false.

All atomic propositions are contingencies. Propositional logic endeavors to find arguments consisting of tautological formulas and to avoid contradictory ones. All formulas validly inferred from tautologies are true. Conversely, all formulas validly inferred from contradictions are false. Truth table analysis can determine whether a formula is a tautology, a contingency, or a contradiction.

2.3.2 Truth Tables

Table 2.3.2-1 shows the truth values for propositions, p and q, and their associated five fundamental connectives. Here, p and q are distinct. They are not the same proposition.

Truth tables for p and q and associated five fundamental connectives ($t = true$, $f = false$).						
p	q	$\sim p$	$p \wedge q$	$p \vee q$	$p \rightarrow q$	$p \leftrightarrow q$
t	t	f	t	t	t	t
f	t	t	f	t	t	f
t	f	f	f	t	f	f
f	f	t	f	f	t	t

Table 2.3.2-1

The entries along the first row of the table are formulas. Truth values are listed in the column beneath. Truth values for p and q take on two values each for a total of four pairs of values:

$$p(v) = t, \quad q(v) = t.$$
$$p(v) = f, \quad q(v) = t.$$
$$p(v) = t, \quad q(v) = f.$$
$$p(v) = f, \quad q(v) = f.$$

In general, if there are n atomic propositions in an argument, there will be 2^n possible truth value combinations. The seven formulas listed in the first row of the table are contingencies, specifically not tautologies or contradictions. If any were tautologies, all the entries in the column below the formula would read t (or f if they were contradictions).

14

2.4 Rules for the Five Fundamental Connectives

Connectives denote a logical operation whereby a new proposition emerges from given propositions. The five fundamental connectives belong to set, C, of Table 2.1.1-1.

2.4.1 Negation

The connective, "~," signifies negation. Its effect is to change the formula's truth value. For any formula p, if p is true, then $\sim p$ is false, and, conversely, if p is false, then $\sim p$ is true. Connectives are generally binary, linking two propositions together, but the connective, ~, is an exception. It operates on a single proposition.

2.4.2 Conjunction

The connective, "∧," called "logical conjunction," combines two formulas into one. It corresponds to the meaning of the English word "and." Hence, if p and q are formulas, then $p \wedge q$ is true if, and only if, both p and q are true; otherwise $p \wedge q$ is false.

For example, since all the truth values listed in the column below the formula, "$j \wedge \sim j$," of Table 2.4.2-1 have a truth value of f, $j \wedge \sim j$ is, therefore, a contradiction. This truth of logic is essential to the theory of proof since the negation of a contradiction is true. Hence, $\sim(j \wedge \sim j)$ is a tautology.

Contradiction and Tautology			
j	*~j*	*j ∧ ~j*	*~(j ∧ ~j)*
t	*f*	*f*	*t*
f	*t*	*f*	*t*

Table 2.4.2-1

2.4.3 Disjunction

The binary connective, "∨," called "disjunction," is true if at least one of the disjuncts is true. While ∨ is often thought to correspond to the

15

English word "or," this is not always the case. The word, "or," is commonly used in the exclusive sense. For instance, the sentence, "I will go to the movies or stay home," implies that I will either go to the movies or stay home, not both. In propositional logic, however, ∨ is inclusive, allowing for the possibility of both options. However, "or," used in the exclusive sense, written in terms of logical connectives, has the formula, "$(p \lor q) \land \sim(p \land q)$," (see Table 2.4.3-1).

Exclusive "or"				
p	q	$p \lor q$	$\sim(p \land q)$	$(p \lor q) \land \sim(p \land q)$
t	t	t	f	f
f	t	t	t	t
t	f	t	t	t
f	f	f	t	f

Table 2.4.3-1

The last column in Table 2.4.3-1 shows that propositions p and q cannot both be true, as expected if "or" is used in the exclusive sense.

2.4.4 Implication

The logical connective, "→," referring to "implication" (see Table 2.4.4-1), is important and also controversial. It is essential because all mathematical proofs are implications, i.e., $p \rightarrow q$. In ordinary language, an implication is an "if, then" statement. For instance, "If I do my homework, then I can watch a movie." Hence, the truth that "I can watch a movie" depends on whether I complete my homework. Logical implication is counterintuitive to the everyday use of "implication." "$p \rightarrow q$" suggests that the truth value of q depends on the truth value of p. However, defined logically, "implication" does not have this property.

A glance at Table 2.1.1-1 shows that the connective, →, does not appear in the set of connectives, C. Its meaning relies on the notion of definition within propositional logic. The rules of propositional logic forbid creating new logical objects that are not already part of the system (i.e., any that do not appear in Table 2.1.1-1). Such connectives

16

provide convenience. They cannot interject further information into the logic system. Hence, $p \rightarrow q$ is defined as $\sim p \vee q$. In short, $p \rightarrow q$ is logically equivalent to $\sim p \vee q$, written as:

$$p \rightarrow q \equiv \sim p \vee q.$$

If the truth values of two or more formulas are the same, then those formulas are "logically equivalent."

Implication				
p	q	$\sim p$	$\sim p \vee q$	$p \rightarrow q$
t	t	f	t	t
f	t	t	t	t
t	f	f	f	f
f	f	t	t	t

Table 2.4.4-1

The truth values for $p \rightarrow q$, displayed in the last column of Table 2.4.4-1, are identical to the truth values of $\sim p \vee q$. Hence:

$$p \rightarrow q \equiv \sim p \vee q.$$

Incidentally, $p \rightarrow p$ is a tautology, and $p \rightarrow \sim p$ is a contingency (see Table 2.4.4-2).

Truth Tables for "$p \rightarrow p$" and "$p \rightarrow \sim p$."			
p	$\sim p$	$p \rightarrow p$	$p \rightarrow \sim p$
t	f	t	f
f	t	t	t

Table 2.4.4-2

In ordinary language, implication suggests a necessary relationship between p (the antecedent) and q (the consequent), implying that if p is true, then q must also be true. For instance, "If I kiss her, she will slap me." The implication is that the kiss will directly cause the slap. Logical implication entails no such causality. Consider the

17

implication, "if you are dead, then you are reading this book." Since you are reading this book, you are not dead. Common sense dictates that such an implication should be false. Let p = "*you are dead*" and q = "*you are reading this book.*" Logically speaking, this instance of the formula, $p \rightarrow q$, is true. You are reading this book. The rules of propositional logic maintain that your being alive, dead, and reading this book is irrelevant. Now consider the implication, $q \rightarrow p$: "If you are reading this book, then you are dead." In this case, $q \rightarrow p$ is false since the antecedent is true, but the consequent is false. Propositional logic addresses these peculiarities through "tautological implication" (discussed in greater detail below). Propositional logic demands that if p and q are atomic and distinct, they must be independent. Hence, the truth value of p cannot depend on the truth value of q, and vice versa.

2.4.5 Bi-Conditional

Finally, the connective, "$\leftrightarrow$," the bi-conditional, is unlisted in Table 2.1.1-1. Its definition is:

$$p \leftrightarrow q \equiv (p \rightarrow q) \wedge (q \rightarrow p).$$

The bi-conditional is true if p and q have the same truth values. Otherwise, it is false. The closest common usage of $\leftrightarrow$ is "sameness" or "equality," but this refers to the bi-conditional truth values, not what the propositions express. The last two columns of Table 2.4.5-1 show that:

$$p \leftrightarrow q \equiv (p \rightarrow q) \wedge (q \rightarrow p).$$

Bi-Conditional					
p	q	$p \rightarrow q$	$q \rightarrow p$	$(p \rightarrow q) \wedge (q \rightarrow p)$	$p \leftrightarrow q$
t	t	t	t	t	t
f	t	t	f	f	f
t	f	f	t	f	f
f	f	t	t	t	t

Table 2.4.5-1

2.4.6 Comments on Connectives

Finally, note that ∧ can be written in terms of ∨, and vice versa:

$$p \wedge q \equiv \sim(\sim p \vee \sim q), \qquad p \vee q \equiv \sim(\sim p \wedge \sim q).$$

These logical equivalences, called De Morgan's Laws, imply that only one of ∧ or ∨ is required, which is correct. That said, it is a standard convention to include both in most treatments of the subject.

Moreover, logical separators indicate which connectives dominate. In the absence of logical separators, the convention is "↔" and "→" dominate "∨" and "∧." However, under this convention, it is not clear what $p \wedge q \vee r$ or $p \leftrightarrow q \rightarrow r$ means. Hence, logical separators must be used appropriately.

2.5 The Meta-Language of Propositional Logic

The set, R (Table 2.1.1-1), contains elements associated with the meta-language. Meta-statements concern formulas and take the following forms: $p \equiv q$, $p = q$, and $p \in P$. The statement, $p \equiv q$, indicates that the formulas, p and q, are logically equivalent, which is tantamount to saying that p and q have identical truth tables. Thus, all atomic propositions are logically equivalent.

Meta-statements using the relational symbols, $\equiv$, $=$, and $\in$, are not formulas in and of themselves but are instead informational statements about formulas.

2.5.1 Contextually Equivalent Propositions

The relational symbol, "$=$," i.e., $p = q$, indicates that p and q are identical contextually. For instance, if

> $p = $ "*Jack does not love Jill,*"
> $q = $ "*It is not the case that Jack loves Jill,*"

then $p = q$. Each proposition, p, q, says the same thing contextually.

2.6 Argument Forms

The formal structure of propositional logic facilitates the study of the anatomy of deductive arguments. A survey of argument forms identifies which forms are valid and which are invalid. The legitimacy of all mathematical proofs rests upon valid arguments.

2.6.1 Valid Argument Forms

Given a set of premises, $p_1, p_2, \ldots, p_n$, and a conclusion, C, then an argument is valid if and only if

$$p_1 \wedge p_2 \wedge \ldots \wedge p_n \to C$$

is a tautology. Otherwise, the argument is a fallacy (invalid).

Note, however, that a valid argument does not guarantee a true conclusion. False premises could lead to a true conclusion, and so a valid argument only guarantees the conclusion is true if the premises are true [177]. For example, consider the argument:

> Either the princess or the queen will attend the ceremony.
> The princess will not attend.
> Therefore, the queen will attend.

The formalization of this argument follows:

$$(p \vee q) \wedge \sim p \to q.$$

p = *"The princess will attend the ceremony."*
q = *"The queen will attend the ceremony."*

Construct a truth table to determine whether the logical form is valid (Table 2.6.1-1).

p	q	$p \vee q$	$\sim p$	$(p \vee q) \wedge \sim p$	$(p \vee q) \wedge \sim p \to q$
t	t	t	f	f	t
t	f	t	f	f	t
f	t	t	t	t	t
f	f	f	t	f	t

Table 2.6.1-1

The table above's last column shows that all the entries below the formula are t's. Hence, "$(p \vee q) \wedge \sim p \to q$" is a tautology. Therefore, the argument form is valid [175].

Consider the argument form:

$$(p \to q) \wedge q \to p.$$

An instance of this form is:

> If the princess attends the ceremony, the queen will attend.
> The queen will attend.
> Therefore, the princess will attend.

p	q	$p \to q$	$(p \to q) \wedge q$	$(p \to q) \wedge q \to p$
t	t	t	t	t
t	f	f	f	t
f	t	t	t	f
f	f	t	f	t

Table 2.6.1-2

The second to last row in this last column of Table 2.6.1-2 has a truth value of f. Hence, $(p \to q) \wedge q \to p$ is a fallacy, as might be expected, since the queen's attendance does not automatically guarantee the princess will attend. With a slight modification (i.e., $(q \to p) \wedge q \to p$), the argument becomes valid. It now reads:

If the queen attends the ceremony, the princess will attend.
The queen will attend.
Therefore, the princess will attend.

A valid argument is one where the conclusion cannot be false if the premises are true. An argument being valid is insufficient to guarantee the truth of its conclusion. Consider the following example:

Either Rembrandt painted the *Mona Lisa,* or Michelangelo did.
Rembrandt did not paint it.
Therefore, Michelangelo did.

The argument is formalized as $(p \lor q) \land \sim p \to q$, which is valid. However, the conclusion is false. The first premise is false; neither Rembrandt nor Michelangelo painted the *Mona Lisa*. A valid argument form only guarantees that, if the premises are true, so is the conclusion. The converse does not hold. If any of the premises are false, the truth or falsity of the conclusion becomes unpredictable. However, if the conclusion is false, then at least one of the premises must be false. This truth of logic applies to Bell's theorem — an essential inference in quantum mechanics.

2.7 Logical Inference

An argument that has a valid logical form and true premises is described as being "sound." With the possible exception of probability and statistics, virtually all mathematical disciplines rely on the concept of a sound argument. Even in probability and statistics, where conclusions are rarely true or false, but only likely so, a remnant of the sound argument concept remains. In mathematics, the premises are called axioms, that is, they are accepted as true. The idea underlying a sound argument is that propositions validly derived from true propositions are true. Such a derivation is called an "inference." A mathematical proof, in turn, is a conclusion drawn from a series of propositions that are either axioms or valid inference rules.

2.7.1 Valid Inferences

Sound arguments contain valid inferences. Table 2.7.1-1 shows a complete list of valid inferences (see proof [290]). Logical inference endeavors to infer true statements from other true statements.

For instance, the law of modus ponens (Table 2.7.1-1) is a valid inference:

$$p \wedge (p \rightarrow q) \rightarrow q.$$

For $p \wedge (p \rightarrow q)$ to be true, both p and $p \rightarrow q$ must be true. If p is true, q cannot be false, else $p \rightarrow q$ would be false. Hence, q is true. Thus, q can be inferred from:

$$p \wedge (p \rightarrow q).$$

The law of modus tollendo tollens, $\sim q \wedge (p \rightarrow q) \rightarrow \sim p$ (Table 2.7.1-1), is also a valid inference. If $\sim q \wedge (p \rightarrow q)$ is true, both $\sim q$ and $p \rightarrow q$ must be true. However, if $\sim q$ is true, then q is false. Hence, p must be false, else $p \rightarrow q$ would be false. Thus, $\sim p$ is true. Therefore, from $\sim q \wedge (p \rightarrow q)$ infer $\sim p$.

Valid Inferences	
Name	Inference
Modus Ponens	$p \wedge (p \rightarrow q) \rightarrow q$
Modus Tollendo Tollens	$\sim q \wedge (p \rightarrow q) \rightarrow \sim p$
Modus Tollendo Ponens	$\sim p \wedge (p \vee q) \rightarrow q$
Law of Simplification	$p \wedge q \rightarrow p$
Law of Adjunction	$p \wedge q \rightarrow p \wedge q$
Law of Hypothetical Syllogism	$(p \rightarrow q) \wedge (q \rightarrow r) \rightarrow (p \rightarrow r)$
Law of Exportation	$[p \wedge q \rightarrow r] \rightarrow [p \rightarrow (q \rightarrow r)]$
Law of Importation	$[p \rightarrow (q \rightarrow r)] \rightarrow [p \wedge q \rightarrow r]$
Law of Absurdity	$[p \rightarrow q \wedge \sim q] \rightarrow \sim p$
Law of Addition	$p \rightarrow p \vee q$

Table 2.7.1-1

Remembering that $p \rightarrow q \equiv \sim p \vee q$, the law of hypothetical syllogism can be written as:

$$(\sim p \vee q) \wedge (\sim q \vee r) \rightarrow (\sim p \vee r).$$

For $(\sim p \vee q) \wedge (\sim q \vee r)$ to be true, both $\sim p \vee q$ and $\sim q \vee r$ must be true. Since q can be true or false, there are two cases: 1) if q is true, then $\sim q$ is false. Hence r is true, else $\sim q \vee r$ would be false. However, if r is true, then $\sim p \vee r$ is true; 2) if q is false, then $\sim p$ is true, else $\sim p \vee q$ would be false. And yet, if $\sim p$ is true, then so is $\sim p \vee r$. Hence, $p \rightarrow r$ follows validly from:

$$(p \rightarrow q) \wedge (q \rightarrow r).$$

The other inferences listed in Table 2.7.1-1 are also valid. It remains to show, though, how valid inferences create sound arguments.

2.7.1.1 Example of a Sound Argument

> *Example: (A Horse Race). If A wins, then either B or C will place. If B places, then A will not win. If D places, then C will not. In fact, A will win. Therefore, D will not place* [178].

Intuitively, this argument is sound since the first sentence says if *A* wins, then *B* or *C* will place. The penultimate premise, "*A* wins," ensures the conclusion. The formalized version of this argument appears in Table 2.7.1.1-1, where the second column shows the justification for each of the eight steps. Lines 1–4 are premises (accepted as true) and indicated by a "P" in the Justification column. Line 5 is inferred from lines 1 and 4 by modus ponens. Line 6 is inferred from 2 and 4 by modus tollendo tollens. Line 7 is inferred from 5 and 6 by modus tollendo ponens. The conclusion, line 8, follows from lines 3 and 7 by modus tollendo tollens.

Proposition	Justification
1) $A \rightarrow B \vee C$	P
2) $B \rightarrow \sim A$	P
3) $D \rightarrow \sim C$	P
4) A	P
5) $B \vee C$	1,4 Modus Ponens
6) $\sim B$	2,4 Modus Tollendo Tollens
7) C	5,6 Modus Tollendo Ponens
8) $\sim D$	3,7 Modus Tollendo Tollens

Table 2.7.1.1-1

The argument is therefore sound. Each statement is either a premise (assumed to be true) or a valid rule of inference.

2.7.1.2 Conditional Proof

Suppose the horse race argument was slightly modified:

> *If A wins, then either B or C will place. If B places, then A will not win. If D places, then C will not. Therefore, if A wins, D will not place [178].*

The conclusion of the argument is now "$A \rightarrow \sim D$." In this slightly modified form, the statement, "A will win," is no longer a premise but part of the conclusion. As a conclusion, $A \rightarrow \sim D$ cannot be a premise. Conclusions of the form, $p \rightarrow q$, are often tricky — or even impossible — to prove without the method of conditional proof (CP):

> *Rule of CP: If a proposition, s, can be derived from a proposition, r, and a set of premises, then $r \rightarrow s$ can be derived from the premises alone.*

The validity of the CP rule rests on the following rationale: Suppose there is a set of premises, $p_1, p_2, p_3, ..., p_n$, and an additional proposition, r. If s can be validly inferred from $p_1, p_2, p_3, ..., p_n$ and r, then $r \rightarrow s$ can be validly inferred from $p_1, p_2, p_3, ..., p_n$, the premises alone. If not, then $r \rightarrow s$ would be false, and therefore, r would be true and s false. Hence, a false conclusion would follow from $p_1, p_2, p_3, \ldots, p_n$ and r (which are all true). However, as s can be validly inferred from $p_1, p_2, p_3, \ldots, p_n$ and r, it must therefore be true, meaning that $r \rightarrow s$ is true [178].

Proposition	Justification
1) $A \rightarrow B \vee C$	P
2) $B \rightarrow \sim A$	P
3) $D \rightarrow \sim C$	P
4) A	H
5) $B \vee C$	1,4 Modus Ponens
6) $\sim B$	2,4 Modus Tollendo Tollens
7) C	5,6 Modus Tollendo Ponens
8) $\sim D$	3,7 Modus Tollendo Tollens
9) $A \rightarrow \sim D$	4,8 CP

Table 2.7.1.2-1

Employing the CP strategy (Table 2.7.1.2-1), line 4 introduces A as a hypothesis (signified by the letter, "H"). The proof then proceeds as before, except that line 9 introduces the CP rule, which says that if A and $\sim D$ are both true, then $A \rightarrow \sim D$ is true.

2.7.1.3 Identifying Inconsistent Premises

The inferential method assumes the premises are true, which might not be the case. A sound argument can show the inconsistencies of a set of premises:

> *If the contract is valid (V), then Horatio is liable (L). If Horatio is liable (L), then he will go bankrupt (B). If the bank will loan him money (M), he will not go bankrupt (~B). In fact, the contract is valid (V), and the bank will loan him money (M) [178].*

Intuitively, the premises are inconsistent (Table 2.7.1.3-1). From line 4, both V and M must be true, so L and $\sim B$ must also be true. However, since ~B is true, then $L \rightarrow B$ would be false. Accordingly, if $V \wedge M$ is true, $L \rightarrow B$ is false. Moreover, if $V \wedge M$ is false, $L \rightarrow B$ is true. The premises are inconsistent. In other words, they cannot all simultaneously be true.

A formalized proof appears in Table 2.7.1.3-1. Lines 1–4 are the premises, and 5–10 are validly inferred. According to the inference method, $B \wedge \sim B$ (line 10) should be true since the premises are assumed to be true, and the remainder of the proof follows from valid inference rules. By the rules of propositional logic, $B \wedge \sim B$ is a contradiction, thus false, which can only be the case if at least one of the premises is false. If all the premises are true, then the conclusion arises as validly inferred and must be true.

Proposition	Justification
1) $V \rightarrow L$	P
2) $L \rightarrow B$	P
3) $M \rightarrow \sim B$	P
4) $V \wedge M$	P
5) V	4 Law of Simplification
6) $V \rightarrow B$	1,2 Law of Hypothetical Syllogism
7) B	5,6 Modus Ponens
8) M	4 Law of Simplification
9) $\sim B$	3,8 Modus Ponens
10) $B \wedge \sim B$	7,9 Law of Adjunction

Table 2.7.1.3-1

If the premises are inconsistent, valid inferences will lead to a contradiction.

2.7.1.4 Reductio ad Absurdum

The contradiction, "$p \wedge \sim p$," suggests another form of proof called "*reductio ad absurdum*" (reduction to absurdity). The proof proceeds by showing that a conclusion is true because its denial would lead to a contradiction, meaning its negation must be true.

Reductio ad absurdum applied to the horse race argument:

> *If A wins, then either B or C will place. If B places, then A will not win. If D places, then C will not. Therefore, if A wins, D will not place* [178].

A formalized argument appears in Table 2.7.1.4-1. The first three lines are identical to the previous proof. To understand line 4, recognize that $A \wedge D \equiv \sim(A \rightarrow \sim D)$ is the negation of the conclusion. Hence, line 4 introduces the denial of the conclusion as a hypothesis. The argument then proceeds through valid inference rules and concludes at line 12, justified by CP, which shows that accepting the hypothesis would lead to a contradiction, thereby making it false. Therefore, through *reductio ad absurdum*, the negation of the hypothesis is true.

Proposition	Justification
(1) $A \rightarrow B \vee C$	P
(2) $B \rightarrow \sim A$	P
(3) $D \rightarrow \sim C$	P
(4) $A \wedge D$	H
(5) A	4 Law of Simplification
(6) D	4 Law of Simplification
(7) $B \vee C$	1,5 Modus Ponens
(8) $\sim C$	3,6 Modus Ponens
(9) B	7,8 Modus Tollendo Ponens
(10) $\sim A$	2,9 Modus Ponens
(11) $A \wedge \sim A$	5,10 Law of Adjunction
(12) $A \wedge D \rightarrow A \wedge \sim A$	4,11 Conditional Proof
(13) $A \rightarrow \sim D$	12 Reductio Ad Absurdum

Table 2.7.1.4-1

2.8 The Algebra of Propositions

Recall that there are only three types of formulas (tautologies, contingencies, and contradictions). Let p, q, r represent contingent formulas, t constitute any tautology and f any contradiction. Once this is accepted, the formulas in Table 2.8-1 become logically equivalent. The equivalences suggest an algebra of propositions similar to numerical algebra [*179*].

The Algebra of Propositions	
Idempotent Laws	
(1) $p \vee p \equiv p$	(1b) $p \wedge p \equiv p$
Associative Laws	
(2) $(p \vee q) \vee r \equiv p \vee (q \vee r)$	(2b) $(p \wedge q) \wedge r \equiv p \wedge (q \wedge r)$
Commutative Laws	
(3) $p \vee q \equiv q \vee p$	(3b) $p \wedge q \equiv q \wedge p$
Distributive Laws	
(4) $p \vee (q \wedge r) \equiv (p \vee q) \wedge (q \vee r)$	(4b) $p \wedge (q \vee r) \equiv (p \wedge q) \vee (q \wedge r)$
Identity Laws	
(5) $p \vee f \equiv p$	(5b) $p \wedge t \equiv p$
(6) $p \vee t \equiv t$	(6b) $p \wedge f \equiv f$
Complementary Laws	
(7) $p \vee \sim p \equiv t$	(7b) $p \wedge \sim p \equiv f$
(8) $\sim\sim p \equiv p$	(8b) $\sim t \equiv f, \sim f \equiv t$
De Morgan's Laws	
(9) $\sim(p \vee q) \equiv \sim p \wedge \sim q$	(9b) $\sim(p \wedge q) \equiv \sim p \vee \sim q$

Table 2.8-1

2.9 Concluding Remarks

Propositional logic comprehensively explores methods with which to join and modify atomic propositions to form more complicated propositions, statements, or sentences. Logical relationships and properties arise from combining or altering statements. The simplest statements (atomic propositions) are indivisible units. Hence, propositional logic studies the logical properties and relations of complete statements rather than their parts, such as subjects or predicates. Propositional logic involves studying logical operators and connectives that produce complex statements whose truth value depends entirely on the simpler statements' truth values. Every statement is assumed to be either true or false, not both [284].

Chapter 3: Predicate Calculus

It is easy to picture the dismay of the innocent person who out of curiosity looks into the later part of the [Principia Mathematica]. He would come upon whole pages without a single word of English below the headline; he would see, instead, scattered in wild profusion, disconnected Greek and Roman letters of every size interspersed with brackets and dots and inverted commas, with arrows and exclamation marks standing on their heads, and with even more fantastic signs for which he would with difficulty so much as find names.

– From the London magazine, **The Spectator**

The previous chapter discussed propositional logic, which emerged from the fundamental notions listed in Table 2.1.1-1. However, such a system's apparatus is insufficient for expressing logical distinctions of the most prominent and elementary kind. The rules of arithmetic fail to materialize within propositional logic. Moreover, there is no means of symbolizing common grammatical distinctions, such as proper nouns, pronouns, adjectives, or adverbs. A richer set of notions is therefore needed to support any group of systematic facts [178]. Such a system is first-order "predicate calculus," also known as "mathematical logic."

Depending on the approach to developing predicate calculus, two nomenclatures arise. Most mathematicians take a somewhat realist approach, where sets or classes comprise fundamental notions within the object language. Logicians generally pursue a more nominalist approach by omitting references to sets or classes and instead employ language forms. For instance, a realist might write "$\forall x(x \in W)$" to indicate that for all x, x belongs to the "class" of white things (W). On the other hand, a nominalist might write "$\forall x(Wx)$" to say for all x, x is white, thereby avoiding any reference to classes. Regardless of which approach is pursued, the syntactical rules that emerge are identical.

New notions within predicate logic complicate the construction of valid inferences. Table 3-1 shows a collection of sets intended to represent an entire description of predicate logic.

Predicate Logic	
Object Language	**Meta-Language**
$T = \{a, b, ..., z\}$	
$P = \{A, B, ..., Z, ...\}$	
$C = \{\wedge, \vee, \sim\}$	
$S = \{(\,), [\,], ...\}$	
$Q = \{\forall, \exists\}$	
$R = \{\in, =, \equiv\}$	

Table 3-1

Predicate calculus' original purpose was to create a wholly self-contained logic system independent of external factors. However, Gödel showed that creating a consistent nontrivial autonomous formal system of logic of this kind was, naturally, impossible. Still, addressing the challenges of creating such a system will be highly instructive.

3.1 Terms and Predicates

Predicate calculus replaces the atomic proposition with two notions:

1. Terms.
2. Predicates.

Terms come in two varieties:

1. Constant.
2. Variable.

Constant terms designate particular things (common or proper nouns). For instance, Socrates, that red pen, or that car in the garage are all constant terms. Variable terms never appear in isolation but rather acquire meaning only when used in conjunction with other notions. Generally, lowercase letters of the alphabet designate terms. Letters at the beginning of the alphabet designate constant terms, while those at the end signify variable terms. However, such conventional designations are arbitrary.

Terms and predicates create sentences or parts of sentences, for example, "*x is white,*" "*x > 0,*" or "*Kermit is a frog.*" "*x*" is a variable term, and "*is white*" is the predicate. In older terminology, "*x is white*" was a "propositional function." Sentences involving constant terms, such as $k \in F$ or Fk, indicate sentences like "*Kermit is a frog,*" where "$k = Kermit$" is the term, and "$F = is\ a\ frog$" is the predicate. "*F*" is a one-place predicate. There can also be two-place predicates, such as "*OHS,*" indicating relationships like "*he is older than she.*" Additionally, there can be *n*-place predicates.

3.1.1 Open Sentences and Free Variables

Forms such as "*x > 0,*" which contain a variable term, are called "open sentences," in which *x* is a free variable. Until a particular value is substituted for the variable, an open sentence is neither true nor false. Since "*x > 0*" is an open sentence, replacing *x* with "*three*" makes 3 > 0 a sentence, since 3 > 0 is true. Only sentences can be assigned a truth value.

3.2 Quantifiers

Within classical propositional logic, assigning truth values to whole classes of objects is far from straightforward. Predicate calculus accomplishes classifications through quantifiers ($\forall, \exists$). For example, "$\forall x(x \in W)$" or "$\forall x(Wx)$" reads, "for all *x*, *x* belongs to the class of white things." If the domain of *x* is all things white, then the sentence, "$\forall x(x \in W)$," can be designated as true. The sentence, "$\exists x(x \in W)$" ("There is at least one 'white thing' in the domain of all things."), can also be deemed true.

The symbol, "∀" (universal quantifier), modifies all terms in a universe of terms, while "∃" (existential quantifier) modifies at least one term.

If a variable in a sentence is quantified, it is "bound." If not, it is "free." Variables found in sentences are bound. Only open sentences contain free variables.

3.2.1 Universes

In propositional logic, the truth of a given atomic proposition depends on verifications outside the system's object language. Ideally, a sentence's validity should be due to its logical form, decided within the object language. Such logical structures exist, but only within a model (i.e., a universe).

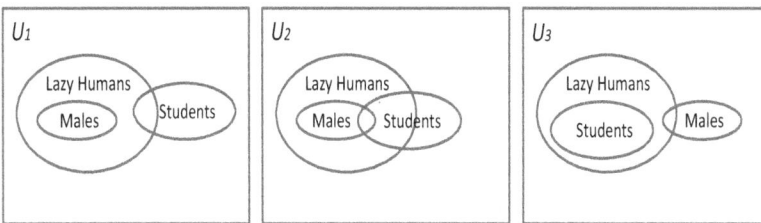

Figure 3.2.1-1

Figure 3.2.1-1 shows three universes, "U_1, U_2, U_3." In all three cases, the universal class (otherwise known as the domain) is all human beings. However, each universe is different. Sentences that are true in one might be false in another. The sentence, "all males are lazy," is true in U_1 and U_2, but false in U_3. The sentence, "Fido is lazy," would be nonsensical since Fido is a dog, and dogs exist outside the domain of the sample universes.

3.3 Formulas in Predicate Calculus

All propositions in propositional logic's object language are formulas. Adopting new notions (terms, predicates, and quantifiers) in predicate calculus raises the question, "what constitutes a formula?" Open sentences, such as "Hx" or "Ax," are formulas and thus considered

34

part of predicate calculus' object language. Expressions like "*Hx*" or "*Ax*," containing a free variable, are not sentences and thus cannot be assigned a truth value. On the other hand, a sentence is a formula with bounded variables, and hence, can be assigned a truth value. Predicate calculus has a slightly different set of syntactic rules than does propositional logic:

1. All open sentences are formulas (*Hx, Fxy*).
2. All sentences are formulas (*Gab*).
3. If ϕ and ψ are formulas, then $\sim\phi$, $\phi \wedge \psi$, $\phi \vee \psi$, $\phi \rightarrow \psi$, and $\phi \leftrightarrow \psi$ are formulas.
4. Any formula quantified by a quantifier ($\forall, \exists$) is a formula.
5. Any formula containing the appropriate use of logical separators is a formula.
6. Other formulas are derived from the above rules.

As with propositional logic, formulas constitute predicate calculus' object language.

3.3.1 Making Sentences

The sentence, "every human is an animal," is an atomic proposition within propositional logic. However, predicate logic breaks the same sentence into smaller pieces and formulates it this way:

$$\forall x(Hx \rightarrow Ax).$$

The sentence can be translated as "for every x, if x is a human, then x is an animal."

In ordinary grammar, "is an animal" is the predicate of "every human is an animal." The predicate logic translation adds the predicate "is a human," which replaces the common noun, "human." "$\forall x(Hx \rightarrow Ax)$" is the standard form for sentences of the type, "every such and such is so and so" [178]. The sentence, "no human is an animal," is formalized as:

$$\forall x(Hx \rightarrow \sim Ax).$$

This can then be translated as "for every x, if x is a human, then x is not an animal."

The existential quantifier, ∃, translates sentences like "some humans are animals" as:

$$\exists x(Hx \wedge Ax).$$

This translates as "there exists at least one x, such that x is a human and x is an animal."

3.3.1.1 Valid Inferences within Predicate Logic

Like propositional logic, predicate logic is primarily concerned with language forms rather than an argument's content. However, both have two critical inferential criteria:

1. The logical inference rules must allow ONLY those conclusions that validly follow from the premises.
2. Logical inferences must permit ALL conclusions which validly follow from the premises [178].

Propositional logic meets both criteria, but predicate calculus can only meet the first in reference to a model. For example, the sentence, "Fido is lazy," is true only if the model's domain includes the subclass, "lazy dogs," and "Fido" is a term within the model. For the sentence, "all dogs are lazy," to be true, "laziness" must be true of every dog in the model.

3.3.1.2 Model Interpretation

The assignment of truth values requires the specification of a model. Once defined, the model's truth values arise for all of the sentences within it. Sentences, restricted to a specific model, satisfy criterion 1. Meeting criterion 2, however, is more problematic. All the conclusions that follow from a set of premises with a valid logical form should be true, suggesting the concept of an "interpretation."

Roughly speaking, an interpretation preserves a model's logical structure. Therefore, if a sentence is true in a particular model, every model interpretation of that sentence is also true. For example, suppose the sentence, "all dogs are lazy," is true within a model's domain. If the domain becomes "all real numbers," then "all rational numbers are real" is an interpretation of "all dogs are lazy" since, within its domain, the sentence, "all rational numbers are real," is true.

There are rules for validly interpreting a sentence. Intuitively, it must be in a one-to-one correspondence. The sentence's form requires preservation. For instance, "some rational numbers are real" is not a valid interpretation of "all dogs are lazy" due to the inherent difference between the words, "some" and "all."

The sentence, "all x are rational," is not a valid interpretation of "all dogs are lazy" since the sentence, "all x are rational," is, unlike "all dogs are lazy," an open sentence. Likewise, "all rational numbers are real" is not a valid interpretation of "all dogs are lazy and hairy" since the predicate, "are real," is a one-place predicate, and the predicate, "lazy and hairy," is a two-place predicate. While certain situations allow for replacing a two-place predicate with a one-place predicate, this type of interpretation is generally invalid.

All terms and predicates should be in the domain of the interpretation. Indeed, it would make no sense to say, "*Jack < Jill*" or, "2 *loves* 3."

Model interpretation within predicate logic provides a framework for describing useful logical concepts, such as "universal validity," "logical implication," and "consistency." An argument is universally valid if and only if every interpretation of it is true in every non-empty universe of terms. The intuitive idea behind this concept is that an argument is true in all possible universes and their model interpretations. A formula, q, logically follows from a formula, p, if $p \rightarrow q$ is universally valid. An argument is consistent if it has at least one true interpretation in some non-empty universe of terms [178].

3.4 Valid Argument Forms within Predicate Calculus

There is no systematic method for determining what constitutes a valid argument within predicate calculus. In propositional logic, truth table analysis, while inefficient, reliably tests an argument's validity. In principle, so long as the argument contains a finite number of premises, a limited number of steps is sufficient for determining validity. However, this was shown not to be possible in all cases by the American logician Alonzo Church (June 14, 1903–August 11, 1995) and the British logician Alan Turing (June 23, 1912–June 7, 1954). Instead, the rules for determining the validity of an argument form rely on trial and error. There are, however, rule-governed procedures for testing validity, which yield an answer in a finite number of steps for some, though not all, argument forms.

Consider the following arguments:

1. All males are lazy.
   Some students are males.
   Therefore, some students are lazy.

2. All students are lazy.
   Some lazy people are males.
   Therefore, some males are students.

3. All students are lazy.
   All lazy people are males.
   Therefore, some males are students.

Which of these three are valid? A valid argument is one in which the conclusion cannot be false if the premises are true. One way of determining whether an argument is invalid is by finding a model universe in which the premises are true, but the conclusion is false. In the model universe, U_3 (Figure 3.2.1-1), the premises of argument 2 are true, but the conclusion is false. The model serves as a counterexample to the claim that argument 2 is valid. Any argument taking 2's form is invalid. A valid argument must be valid in all universes and their model interpretations.

What about argument 3? Examining the three universes (U_1, U_2, and U_3), none provide a counterexample showing that argument 3 is invalid. And yet, moving the male oval in U_3 entirely inside the lazy humans oval to stop it intersecting with the students oval provides the counterexample, which reveals argument 3 to be an invalid argument form.

At this point, the only way of determining argument 1's validity is to check every interpretation of the corresponding model universes, possible only if the number of model interpretations is finite. However, there is no indication that the number of models, or their interpretations, is bounded in this case. There is no bound on the number of model variants associated with a given argument form in predicate logic. Other rule-based methods (developed subsequently) will show that argument 1 is a valid argument form [175].

3.4.1 Developing Valid Arguments

Most proofs within predicate calculus require four steps:

1. Symbolize the premises into the object language.
2. Drop quantifiers following specific rules.
3. Apply valid inferences.
4. Add quantifiers to obtain the conclusion [178].

3.4.1.1 Step 1: Symbolizing

Symbolizing premises requires a great deal of ingenuity and practice. Translating ordinary language into predicate calculus' object language can be difficult to the point of impossible. Nevertheless, with practice, most sentences translate into the object language. Moreover, predicate calculus provides a richer set of sentential capabilities than does propositional logic.

3.4.1.2 Steps 2–4: Dropping and Adding Quantifiers

Predicate calculus comes with a more complicated set of inference rules than propositional logic. The new rules primarily involve specifying when and if quantifiers can be dropped or restored, providing strategies on applying valid inferences, and suggesting how to arrive at a valid conclusion. Additionally, the rules, developed primarily through trial and error, prevent invalid inferences.

3.4.2 Universal Specification and Generalization

Example (from Lewis Carroll): No ducks are willing to waltz. No officers are unwilling to waltz. All my poultry are ducks. Therefore, none of my poultry are officers [178].

Sentence	Justification
1) $\forall x(Dx \rightarrow \sim Wx)$	P
2) $\forall x(Ox \rightarrow Wx)$	P
3) $\forall x(Px \rightarrow Dx)$	P
4) $Dx \rightarrow \sim Wx$	1, US
5) $Ox \rightarrow Wx$	2, US
6) $Px \rightarrow Dx$	3, US
7) Px	x H
8) Dx	x 6,7 Modus Ponens
9) $\sim Wx$	x 4,8 Modus Ponens
10) $\sim Ox$	x 5,9 Modus Tollendo Tollens
11) $Px \rightarrow \sim Ox$	7,10 CP
12) $\forall x(Px \rightarrow \sim Ox)$	11, UG

Table 3.4.2-1

The first step is to translate the premises of Carroll's argument into the object language, accomplished in lines 1–3 of the above table. Lines 4–6 illustrate the first rule involving quantifiers, universal specification (US). A universal quantifier should be dropped if it runs through the entire scope of the formula, justified because whatever is true of every term is true for any given term. Line 7 introduces the hypothesis,

"Px." No truth value can be assigned to a premise containing a free variable. Indiscriminately introducing premises can lead to invalid inferences, such as in *"Px → ∀x(Px)."* Concluding that everything is poultry from the premise that something is poultry would be a fallacious inference. Introducing a premise containing a free variable is valid in CPs. In lines 7–10, the Justification column contains an *x*, which signifies that no universal generalization (UG) is allowed on the free variable, *x*, until line 11, where the restriction is removed, justified by CP. Lines 8–10 follow from valid inferences. The final rule, UG, allows the reinstatement of the universal quantifier so long as no free variables appear in the formula universally quantified — accomplished in line 12.

3.4.3 Existential Specification and Generalization

Example: All mammals are animals. Some mammals are two-legged. Therefore, some animals are two-legged [178].

Sentence	Justification
1) $\forall x(Mx \rightarrow Ax)$	P
2) $\exists x(Mx \land Tx)$	P
3) $M\alpha \land T\alpha$	2, ES
4) $M\alpha \rightarrow A\alpha$	1, US
5) $M\alpha$	3, Law of Simplification
6) $A\alpha$	4,5 Modus Ponens
7) $T\alpha$	3, Law of Simplification
8) $A\alpha \land T\alpha$	6,7 Law of Adjunction
9) $\exists x(Ax \land Tx)$	7, EG

Table 3.4.3-1

In line 3 (Table 3.4.3-1), the new rule of existential specification (ES) allows for stripping off the existential quantifier. The justification for this is that if something satisfies a condition, then some namable individual meets it. The symbol, "α," meaning ambiguous name, represents a namable individual and is usually specified using lowercase Greek letters. Line 4 follows from line 1 by US. Lines 5–8

41

are valid inferences. Finally, line 9 introduces existential generalization (EG), which justifies the conclusion since, if a named individual satisfies a condition, at least one individual satisfies that condition, i.e., there exists something that is both an animal and two-legged.

3.4.3.1 Rules for Using Ambiguous Names

However, the use of ambiguous names can create a subtle fallacy, avoidable if line 3 is placed before line 4 in the previous argument [*178*] (Table 3.4.3.1-1).

An ambiguous name used with one premise (expressing one condition) cannot be used again with another premise (expressing another condition). An individual meeting a specific condition does not imply that the same individual meets a second condition. Therefore, always apply ES before US when using the same ambiguous name.

Sentence	Justification
1) $\exists x(Hx)$	P
2) $\exists x(\sim Hx)$	P
3) $H\alpha$	1, ES
4) $\sim H\alpha$	2, ES (Fallaciously)
5) $H\alpha \wedge \sim H\alpha$	3,4 Law of Adjunction
6) $\exists x(Hx \wedge \sim Hx)$	7, EG

Table 3.4.3.1-1

3.4.4 Additional Rules of Inference

A few additional rules govern the removal and inclusion of quantifiers to avoid fallacies. A truth of arithmetic says that there is no largest integer, written "$\forall x \exists y(x < y)$." Applying US, i.e., $\exists y(x < y)$, then ES, the previous open sentence becomes $x < \alpha$, after which applying EG leaves:

$$\exists x(x < x).$$

This formula is clearly fallacious as no number is less than itself. The problem arises in step two, i.e., $\exists y(x < y)$. The variable, x, which is free, is not in the scope of the quantifier, $\exists$. Applying EG, or any generalized quantifier, to a variable outside the quantifier's scope is incorrect. Such a fallacy can be avoided by sub-scripting the ambiguous name, i.e., "$x < \alpha_x$" with the same letter as the free variable, and then avoiding the application of EG to an ambiguous subscripted name. The same holds for UG. From $x < \alpha_x$, it does not follow that $\forall x(x < \alpha_x)$ and then $\exists y \forall x(x < y)$, which says that y is the greatest integer, which is false.

Example: *All horses are animals. Therefore, all heads of horses are heads of animals* [178].

Sentence	Justification
1) $\forall x(Px \rightarrow Ax)$	P
2) $\exists y(Py \wedge Hxy)$	xP
3) $P\alpha_x \wedge Hx\alpha_x$	x2, ES
4) $P\alpha_x \rightarrow A\alpha_x$	1, US
5) $A\alpha_x \wedge Hx\alpha_x$	x,3,4 T
6) $\exists(y)(Ay \wedge Hxy)$	x,5 EG
7) $\exists y(Py \wedge Hxy) \rightarrow \exists y(Ay \wedge Hxy)$	2,6 CP
8) $\forall(x)[\exists y(Py \wedge Hxy) \rightarrow \exists y(Ay \wedge Hxy)]$	7, UG

Table 3.4.4-1

Line 2 introduces the premise, "$\exists y(Py \wedge Hxy)$." There is no restriction on adding premises, but note that this premise contains a free variable (x), indicated by placing an x in the Justification column. Since x is free, the restrictions noted in Table 3.4.4-1 apply. Introducing a premise requires CP. Line 3 follows from existential specification by replacing y with the ambiguous name, α, which is subscripted with an x to prevent invalid quantification involving x. Line 4 follows from line 1 by US. Line 4 follows 3, thereby permitting the use of the same ambiguous name in both lines. The subscripted x carries along. Line

5 is a valid inference. Existential generalization in line 6 quantifies y. Since y is not free, it quantifies. Line 7 follows the CP rule, which allows the restriction on x to be dropped. Hence, the conclusion in line 8 follows by UG. For more information on the properties of quantifiers, see the Appendix.

3.4.5 Summary of the Rules of Inference

Rules of Inference [178]		
Abbreviation	**Rule**	**Restriction**
P	Introduction of Premises	None.
T	Valid Inferences	None.
CP	Conditional Proof	None.
US	Universal Specification: from $\forall x(S(x))$ derive $S(x)$	x cannot be free.
UG	Universal Generalization: from $S(x)$ derive $\forall x(S(x))$	x cannot be free. x cannot be a free variable used in conjunction with an ambiguous name.
ES	Existential Specification: from $\exists x(S(x))$ derive $S(\alpha)$	The ambiguous name, "α," has not been used previously.
EG	Existential Generalization: from $S(\alpha)$ derive $\exists x(S(x))$	x cannot be free. The x associated with α — an ambiguous name — cannot be free. The ambiguous name, "α," cannot be replaced if it previously occurs within the scope of x. Hence, from $\forall x(S(\alpha))$ derive $\exists x(S(x))$ is invalid.

3.5 Predicate Calculus with Identity

Generally speaking, switching terms and predicates in a sentence results in losing meaning. For instance, "salt is white" is meaningful, but "white is salt" is not. In some cases, however, the sentences do retain their meaning. For example, "James Madison was the fourth president of the United States." Replacing the term with the predicate results in the equivalent statement, "the fourth president of the United States was James Madison." Such a sentence is called an "identity." Identities are sentences where the term and predicate are identical.

Formally, a sentence is an identity if it is:

1. Reflexive: $x \to x$.
2. Symmetric: if $x \to y$, then $y \to x$.
3. Transitive: if $x \to y$ and $y \to z$, then $x \to z$.

Few human relationships are identities: "Is the brother of," "is taller than," and "is the girlfriend of" are not identity relationships. The relation, "is the brother of," satisfies 2 and 3, but not 1. No one is their brother.

3.5.1 Rules Involving Identities

Predicate calculus restricts identities to open sentences. From $a = a$, it follows that $\forall x(x = x)$, or from $a = a$ derive $\exists x(a = x)$. The first follows because all things are equal to themselves, and the second follows because if something exists, something is equal to it (namely, itself). However, deriving $\exists(x)\sim(x = x)$ from $x = y$ and $\exists(x)\sim(x = y)$ is fallacious, confirming the restriction to open formulas [178].

1. An ambiguous name in a previous derivation cannot appear in a subsequent derivation.

Suppose, in some proof, $\exists x(Fx \to \sim F\alpha)$. Now imagine a second proof where $\exists(x)\sim(Fx)$ leads to $\sim F\alpha$ by ES. The use of $\exists x(Fx \to \sim F\alpha)$ from the previous proof is invalid [178]. Each proof is independent, and the restriction on ambiguous names still holds. Again, an individual that meets a specific condition does not guarantee their meeting a second condition.

3.5.2 The Predicate Calculus and Mathematics

Predicate calculus with identity can support the mathematical development of the most common systems. Of course, some valid arguments are unspecifiable in predicate calculus, but these cases are quickly resolved by introducing additional rules such as identity. The axioms of the rational and real numbers written in terms of predicate calculus with identity are displayed in Table 3.5.2-1. They include the non-logical symbols "+," " ⋅," and "<."

The Axioms of the Rational and Real Numbers[2]
1. $\forall x \forall y (x + y = y + x)$
2. $\forall x \forall y (x \cdot y = y \cdot x)$
3. $\forall x \forall y \forall z ((x + y) + z = x + (x + z))$
4. $\forall x \forall y \forall z ((x \cdot y) \cdot z = x \cdot (y \cdot z))$
5. $\forall x \forall y \forall z (x \cdot (y + z) = (x \cdot y) + (x \cdot z))$
6. $\forall x (x + 0 = x)$
7. $\forall x (x \cdot 1 = x)$
8. $\forall x \exists y (x + y = 0)$
9. $\forall x \forall y \forall z (y \neq 0 \rightarrow \exists! z (x = y \cdot z))$
10. $\forall x \forall y (x < y \rightarrow \sim(y < x))$
11. $\forall x \forall y \forall z (x < y \land y < z \rightarrow x < z)$
12. $\forall x \forall y (x \neq y \rightarrow x < y \lor y < x)$
13. $\forall x \forall y \forall z (y < z \rightarrow x + y < x + z)$
14. $\forall x \forall y \forall z (0 < x \land y < z \rightarrow x \cdot y < x \cdot z)$
15. $0 \neq 1$

Table 3.5.2-1

---

[2] Note that the logical symbol "↔" has been replaced by "=," as is customary.

3.6 Concluding Remarks

First-order predicate logic used in mathematics, philosophy, linguistics, and computer science can be distinguished from propositional logic through quantified variables. First-order predicate logic is standard for formalizing mathematics into axioms. Mathematical systems, such as number and set theory, are formalized into first-order axiom schemata, such as Peano arithmetic and Zermelo–Fraenkel set theory.

However, no first-order theory can fully describe structures with an infinite domain, such as the natural numbers or the real line. Systems for these structures appear in stronger logics, such as second-order logic. In first-order theories, predicates associate with sets. In higher-order theories, predicates are sets of sets. Higher-order logics are termed "categorical" or the "theory of types." Russell developed the theory of types to solve emergent contradictions, such as Russell's Paradox.

The axioms of the rational and real numbers have been wholly formalized in Table 3.5.2-1 [*178*]. All theorems of the rational and real numbers follow from these axioms. However, formal proofs tend to be long and tedious. Mathematicians, therefore, use informal proofs instead, which usually involve removing the quantifiers and omitting specific steps. For instance, formally proving the simple statement, "$\forall x(x \cdot 0 = 0)$," requires approximately eight steps. Informally, it requires three. However, turning a formal proof into an informal one can lead to errors, which are not always readily apparent.

Chapter 4: Logic and Set Theory

The Realist position is probably the one which most mathematicians would prefer to take. It is not until he becomes aware of some of the difficulties in set theory that he would even begin to question it. If these difficulties particularly upset him, he will rush to the shelter of Formalism, while his normal position will be somewhere between the two, trying to enjoy the best of two worlds.

— Paul Cohen

It is virtually impossible to discuss even the rudiments of mathematics without the notion of a "set." Set theory emerged in 1873 upon Cantor's revelation that the linear continuum (i.e., the real line) is not countable. In other words, its points cannot be tallied up using natural numbers. Accordingly, although the natural numbers and the real numbers contain an infinite number of elements, there are more real numbers than natural numbers. This unmasking allowed for the investigation into different sizes of infinities [256].

The Cantorian conception of a set defines it as "a collection of objects into a whole of our intuition or our thought," denoted "$S = \{s\}$," where lowercase letters signify the elements or members of S, i.e., $s \in S$.

The collection, "$\{dog, tree, human\}$," is a set but of a type rarely found in practice. More often, the elements of a set share one or more associated properties, for instance, the set of "real numbers" or the set of "students at a university." The real numbers have specific properties, and a university student may have other properties beyond being enrolled at a university.

Cantor failed to clearly differentiate between the elements of a set with associated properties and one where properties are not specified. In axiomatic set theory, the distinction is more evident. Sets associated with properties are typically referred to as "classes" and are signified by "$\{x: \Theta\}$," which specifies that the "$x's$" (elements of a set) have the property, "Θ." On the other hand, a set with only the elements listed is signified as "$\{x\}$" with no reference to a property.

4.1 Set and Logic Relationships

Venn diagrams illustrate set relationships and operations (Figure 4.1-1). The set, U, represents the universe of objects under consideration. The circles labeled "$A, B,$ and C" are divided into distinct areas marked "$a1, a2, a3, ...$" $A, B, C \subset U$, and

$$\forall i(ai \subset U, \qquad i = 1,2,...,8).$$

Some of the $ai's$ are subsets of $A, B,$ or C. For example, $a1 \subset A$, but $a1 \not\subset B, C; a2 \subset A, B,$ but $a2 \not\subset C$.

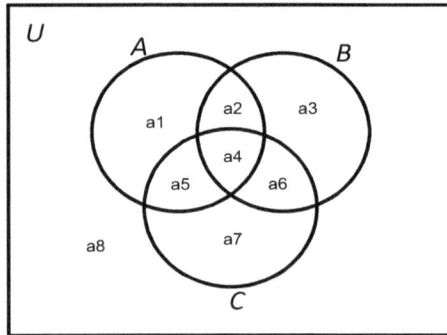

Figure 4.1-1

Classical propositional logic says that an argument is valid when its conclusion cannot be false if its premises are true. For example, Bell's inequality is an exercise in logic, which reads:

$$N(A, {\sim}B) + N(B, {\sim}C) \geq N(A, {\sim}C).$$

$N(A, {\sim}B)$ and $N(B, {\sim}C)$ are the premises, and $N(A, {\sim}C)$ acts as the conclusion. First,

$N(A, \sim B) =$
"the number of elements in a1 plus the number of elements in a5."

In other words, this refers to the number of elements in set, A, but not in B (Figure 4.1-1).

Likewise,

$N(B, \sim C) =$
"the number of elements in a2 plus the number of elements in a3,"

which is the number of elements in set, B, but not in C.

Finally, the conclusion:

$N(A, \sim C)$
$=$ *"the number of elements in a1 plus the number of elements in a2,"*

which is the number of elements in set, A, but not in C.

Hence, Bell's inequality:

$$N(A, \sim B) + N(B, \sim C) \geq N(A, \sim C),$$

says that the number of elements in the set, A, but not in B, plus the number of elements in the set, B, but not in C must be greater than, or equal to, the number of elements in the set, A, but not in C. Substituting the Venn diagram values for their equivalent symbols in Bell's inequality would leave:

$$N(a1 + a5) + N(a2 + a3) \geq N(a1 + a2).$$

The inequality is logically valid since its conclusion (right) cannot be false if the premises (left) are true. In other words, the conclusion is entailed in the premises. This implies that set theory is, in some sense at least, related to logic.

4.2 Well-Defined Sets

All sets in naïve set theory are "well defined," meaning that it is always possible to determine whether an element belongs to a particular set or not. Two sets are equal if they have the same elements.

4.2.1 Set Specifications

If s is an element of the set, S, this is denoted by "$s \in S$." By contrast, "$s \notin S$" signifies that s is not an element of S. An element either belongs to a set, or it does not. Sets can be specified by listing the elements, for example, $A = \{a, b, c\}$, or by logical specification, i.e., $\forall x(x \in B \rightarrow P(x))$, which states that all the x's belonging to B have property, P. Sets can be finite or infinite depending on whether their number of elements is finite or infinite. The elements of a set, S, can themselves be sets, denoted by "$S = \{\{x\}\}$" (read S is a set of sets). For instance, if $A = \{a, b, c\}$, then let "$P(A)$" be defined as:

$$P(A) = \{A, \{a, b\}, \{a, c\}, \{b, c\}, \{a\}, \{b\}, \{c\}, \emptyset\}.$$

$P(A)$ is called the "power set" of A. All the elements of $P(A)$ are sets. In fact, $P(A)$ is the set of all the subsets of A. A set of sets is occasionally referred to as a "family."

4.2.2 Element and Set Associations

"$\in$" associates elements with sets, and "$\subset$" associates sets with sets. Hence, if $A = \{a, b\}$, then $a \in A$ and $\{a\} \subset A$ are meaningful associations, while $\{a\} \in A$ and $a \subset A$ are not. Further still, they would, in fact, be incorrect. If

$$A = \{\{a\}, \{b\}\},$$

then $\{a\} \in A$ is correct, but $\{a\} \subset A$ is not. The correct form is $\{\{a\}\} \subset A$. While $a \in \{a\}$ is correct, $a \in A$ is not.

If $H \subset U$, this does not exclude the possibility that $H = U$, the definition of which is:

$$\forall H \forall U (H \subset U \wedge U \subset H \leftrightarrow U = H) \equiv \forall x (x \in U \leftrightarrow x \in H).$$

The following rules display the relationship between logic and set theory:

1. $\forall x (x \in A \leftrightarrow A \subset A).$
2. $\forall x ((x \in B \rightarrow x \in A \wedge x \in A \rightarrow x \in B) \leftrightarrow A = B).$
3. $\forall A \forall B \forall C (A \subset B \wedge B \subset C \rightarrow A \subset C).$

Note that the three rules above define an identity. In other words, in some sense, set theory is equivalent to logic, in particular, predicate calculus.

4.2.3 Complimentary Sets

The logical connective, "$\sim$," supports additional, complementary, set relationships. For instance, $\sim A$ (sometimes designated A^c or A') is called the "compliment" of A, logically defined as:

$$\sim A \equiv \forall x (x \in U \wedge A \subset U \rightarrow x \notin A).$$

The difference of two sets, $A \wedge \sim B$, is logically defined as:

$$A \wedge \sim B \equiv \forall x (x \in A \rightarrow x \notin B).$$

4.2.4 Classes

Classes (sets with properties) are an essential feature of logical arguments. For example, the validity of the classic syllogism:

> All humans are mortal.
> Socrates is human.
> Therefore, Socrates is mortal,

depends on the idea of a sub-class or subset. The universal class — which in this case is all things mortal — can be designated logically:

$$\forall x\big(x \in U \to M(x)\big), \qquad M(x) = \text{"x is mortal."}$$

All the members of U share the property of being mortal. A sub-class, H, of U, denoted by $H \subset U$, inherits the property of mortality from U. Therefore,

$$\forall x\big(x \in U \wedge x \in H \subset U \to H(x) \wedge M(x)\big), \qquad H(x) = \text{"x is human,"}$$

which says that, if $x \in U, H$, and $H \subset U$, then in addition to the property of being human, from U, x inherits the property of mortality. Substituting Socrates for x creates the sentence "Socrates is both human and mortal." U is a superset of H, written as "$U \supset H$," i.e.,

$$U \supset H \equiv \forall x(x \in H \to x \in U).$$

4.2.5 Universal Set and the Null (Empty) Set

The term universal set or class suggests only one universal set. However, naïve set theory allows for the possibility of many universal sets—university students, frogs in a pond, citizens of a city, a collection of numbers, and so on. All can be universal sets. However, a universal set restricts its domain to those specific classes. Having a universal set results in Russell's paradox. Consider the set of all sets not members of themselves:

$$U = \{U_1, U_2, U_3, U_4, \dots\}.$$

"U" should be a member of itself since it is a set. This becomes problematic as U would then be both a member and not a member of itself—a contradiction. In axiomatic set theory, there is no universal set.

The null (or empty set), denoted by "Ø," is a set without elements, which raises philosophical issues. Does Ø exist, or is it merely a property of a set? It is both. For instance, suppose that:

$$A = \{\exists x(x \in \{1,3,5, \dots\} \to x^2 = 4)\}.$$

This implication is false. There is no $x \in \{1,3,5,...\}$ such that $x^2 = 4$. Hence, $A = \emptyset$. Yet, the implication

$$\exists B(B = \{\emptyset\} \rightarrow B \neq \emptyset)$$

is true. Hence, $\emptyset$ exists and is a set, not simply one's property. But, these difficulties aside, it follows that:

$$\forall A \exists! U(A \subset U \rightarrow \emptyset \subset A \subset U).$$

The null set is considered finite and a subset of every set.

4.3 Set Operations

The symbols, "∪" (union) and "∩" (intersection), denote set operations, which can be defined logically:

1. $A \cup B \equiv \forall x(x \in A \lor x \in B)$.
2. $A \cap B \equiv \forall x(x \in A \land x \in B)$.
3. $\emptyset = A \cap B \equiv \forall x((x \in A \rightarrow x \notin B) \land (x \in B \rightarrow x \notin A))$.

If $A \cap B = \emptyset$, then A and B are deemed "disjoint" sets. They do not have common elements.

All sets within a universe of terms are subsets of U — the universal set. With this information, the concept of being "well defined" can be illustrated using set operations. For instance, in Figure 4.1-1:

1. $a1 \cup a2 \cup ... \cup a8 = U$.
2. $\forall i \forall j(i \neq j \rightarrow ai \cap aj = \emptyset, \ i,j = 1,...,8)$.

The sets, "$a1, a2, ..., a8$," are pairwise disjoint sets that partition U.

The elements belonging to $a1, a2, ..., a8$ equal the total number of elements in U. Conditions 1 and 2 specify a well-defined universe since the elements contained in a set are either in a particular set or are absent from it. Well-definedness allows the sets, A, B, and C and their unions and intersections, to be specified by the pairwise disjoint sets, $a1, a2, ..., a7$.

Using the concept of being well defined, the following relationships can be verified:

1. $A = a1 \cup a2 \cup a4 \cup a5$.
2. $B = a2 \cup a3 \cup a4 \cup a6$.
3. $C = a4 \cup a5 \cup a6 \cup a7$.
4. $A \cap B = a2 \cup a4$.
5. $A \cap C = a4 \cup a5$.
6. $B \cap C = a4 \cup a6$.
7. $A \cap B \cap C = a4$.
8. $A \cup B = a1 \cup a2 \cup a3 \cup a4 \cup a5 \cup a6$.
9. $A \cup C = a1 \cup a2 \cup a4 \cup a5 \cup a6 \cup a7$.
10. $B \cup C = a2 \cup a3 \cup a4 \cup a5 \cup a6 \cup a7$.
11. $A \cup B \cup C = a1 \cup a2 \cup a3 \cup a4 \cup a5 \cup a6 \cup a7$.

4.4 Comparing Set Theory to Logic

Comparing Table 2.8-1 with Table 4.4-1 (below), replacing $\cup$ in the latter by $\vee$, $\cap$ by $\wedge$, $\emptyset$ by f, U by t, and "A, B, C" by "p, q, r," would result in the latter looking identical to the former. This would imply that the algebra of sets is identical to that of classical logic's propositional algebra. If a given mathematical structure, when compared to another, is found to be identical (other than labeling), they are described as isomorphic.

The Algebra of Sets	
Idempotent Laws	
1a. $A \cup A = A$	1b. $A \cap A = A$
Associative Laws	
2a. $(A \cup B) \cup C = A \cup (B \cup C)$	2b. $(A \cap B) \cap C = A \cap (B \cap C)$
Commutative Laws	
3a. $A \cup B = B \cup A$	3b. $A \cap B = B \cap A$
Distributive Laws	
4a. $A \cup (B \cap C) = (A \cup B) \cap (A \cup C)$	4b. $A \cap (B \cup C) = (A \cap B) \cup (A \cap C)$
Identity Laws	
5a. $A \cup \emptyset = A$	5b. $A \cap U = A$
6a. $A \cup U = U$	6b. $A \cap \emptyset = \emptyset$
Compliment Laws	
7a. $A \cup {\sim}A = U$	7b. $A \cap {\sim}A = \emptyset$
8a. ${\sim}{\sim}A = A$	8b. ${\sim}U = \emptyset, \ {\sim}\emptyset = U$
De Morgan's Laws	
9a. ${\sim}(A \cup B) = {\sim}A \cap {\sim}B$	9b. ${\sim}(A \cap B) = {\sim}A \cup {\sim}B$

Table 4.4-1

Using the laws listed in Table 2.8-1, it follows that:

$$\forall A \forall B(A \wedge {\sim}B \rightarrow (A \wedge {\sim}B) \cap B = \emptyset).$$

To see this, since set operations and logical propositional laws are associative,

$$(A \wedge {\sim}B) \cap B = A \wedge ({\sim}B \cap B) = A \wedge \emptyset \equiv A \cap \emptyset = \emptyset.$$

There are various ways of writing "$A \subset B$"[3] logically:

1. $\forall A \forall B (A \subset B \leftrightarrow A \cap B = A)$.
2. $\forall A \forall B (A \subset B \leftrightarrow A \cup B = B)$.
3. $\forall A \forall B (A \subset B \leftrightarrow A \cap {\sim}B = \emptyset)$.
4. $\forall A \forall B (A \subset B \leftrightarrow {\sim}B \subset {\sim}A)$.
5. $\forall A \forall B (A \subset B \leftrightarrow {\sim}A \cup B = U)$.

To see that

$$\forall A \forall B (A \subset B \leftrightarrow A \cap B = A),$$

if $A \subset B$, then

$$\forall x (x \in A \rightarrow x \in B).$$

Hence,

$$x \in A \cap B \rightarrow A \subset A \cap B.$$

However, since

$$\forall x (x \in B),$$

then $A \cap B \subset A$. By the definition of equal sets, $A \cap B = A$.

Conversely, if $A \cap B = A$, then

$$\forall x (x \in A \rightarrow x \in B),$$

---

[3] A note on notation: In logic, the meta-statements "$p = q$" and "$p \equiv q$" mean different things. The former says that p and q say the same thing, while the latter states that p and q are logically equivalent. All atomic propositions are logically equivalent, but they do not say the same thing. In set theory, distinguishing between $p = q$ and $p \equiv q$ is unnecessary since all elements of a set are assumed to be distinct. If it is true that $p, q, r \in A$ and $p = q = r$, then $A = \{p\}$. Here I use the convention that, when relating two set theory statements, "$=$" is used, and when relating a set theory statement to a logical statement, "$\equiv$" is used.

and

$$x \in A \cap B \to A \subset A \cap B.$$

Although,

$$x \in B \to A \cap B \subset B.$$

By transitivity, $A \subset B$. The other four equivalencies above can be proven similarly.

By employing well-defined universes, most sentences within predicate calculus formulate within the language of sets. For instance:

> $U =$ "*Students at a University.*"
> $A =$ "*male student.*"
> $B =$ "*A student whose height is greater than 5'8".*"
> $C =$ "*A student with blue eyes.*"

Sentences written in the language of predicate calculus can be made equivalent to statements written in set theoretical language. For example:

$$A \subset U \equiv \forall x (x \in A \to x \in U), \qquad \exists x (Ax \wedge Cx) \equiv A \cap C \neq \emptyset.$$

The first sentence says that if x is a male student, then x is a university student. The second sentence says, "there exists a male student with blue eyes."

4.5 Concluding Remarks

Just as the laws of logic allow undertaking algebra with logical formulas, the laws of set theory allow undertaking algebra with sets. Due to the close relationship between logic and set theory, their algebras are similar. More precisely, considering only subsets of some given universal set, U, there is a direct correspondence between the basic symbols and operations of propositional logic and certain symbols and operations in set theory (Table 4.5-1).

Logic	Set Theory
t	U
f	$\emptyset$
$p \wedge q$	$A \cap B$
$p \vee q$	$A \cup B$
$\sim p$	$\sim A$

Table 4.5-1

Any valid logical formula or computation involving propositional variables, and the symbols, "t, f, $\wedge$, $\vee$" and "$\sim$," can be transformed into a valid formula or computation in set theory by replacing the formula's propositions with subsets of U, and replacing the logical symbols with "U, $\emptyset$, $\cap$, $\cup$" and the complement operator, "$\sim$" [285].

Chapter 5: The Theory of Realist Logic

The pessimist complains about the wind; the optimist expects it to change; the realist adjusts the sails.

— *William Arthur Ward*

The validity of the logical systems discussed in Chapters 2 and 3 relied on the notion of a sound argument. However, such systems are highly nominalist because logic and mathematics spring from invented formal linguistic rules expressed as language forms.

The realist, on the other hand, demands an independent logical form. However, modern logic lacks such a structure, which is required in a truly realist logic system. The argument, "all $S's$ are $M's$, all $M's$ are $P's$, therefore all $S's$ are $P's$," is simply a linguistic form. If the $S's$, $M's$, and $P's$ need instantiation by independent states of affairs, then current logic systems lack a separate language form representing this.

Logical realism asserts that logical entities exist in an independent mind. Thus, mathematics is discovered, not invented. Furthermore, this perspective suggests that mathematics consists of discoverable objects that exist independently from the ideas of those objects expressed through a language form. In other words, as mathematical objects are discovered, a realist logic system requires two language forms. The first form represents discoverable objects, while the second represents the ideas of a "discoverer."

Thus, this chapter develops a realist logic approach, in which abstract objects — expressed in a language — are instantiated by an independent language.

5.1 The Foundations of Logical Systems

Systems of logic generally share a few common characteristics:

1. **An Object Language**: The philology of the logic system.
2. **Valid/Invalid Forms**: Semantics associated with the object language.
3. **Consistency/Inconsistency**: Compatible or incompatible syntax or semantics.
4. **Completeness/Incompleteness**: Provable or unprovable statements within the object language.

5.2 The Syntax and Semantics of Realist Logic

The set, S, known as the nominal set, is a collection of elements, s, into a whole, S, of intuition or thought, designated "$S = \{s\}$." The instantiation set, $\bar{S}$, being independent of S, is a collection of elements, $\bar{s}$, into a whole designated "$\bar{S} = \{\bar{s}\}$." The syntax of realist logic relates S to $\bar{S}$.

If $s \in S$, then s is an expression of thought. If $\bar{s} \in \bar{S}$, then $\bar{s}$ represents a possible instantiation of the thought in S. Hence,

$$\forall s \forall \bar{s}(s \in S \wedge \bar{s} \in \bar{S} \leftrightarrow \bar{s} \notin S \wedge s \notin \bar{S}).$$

A set, S, is an entity "undivided," i.e., S is instantiated by the "undividedness" of $\bar{S}$ which is a universal concept. The connection between S and $\bar{S}$ is defined by a relationship, r, that associates the elements in S with the elements in $\bar{S}$, designated "$S \xrightarrow{r} \bar{S}$." Moreover, a relationship, $\bar{r}$, associates the elements in $\bar{S}$ with the elements in S and is designated "$\bar{S} \xrightarrow{\bar{r}} S$." Note that $\bar{r}$ instantiates r.

Example (Figure 5.2-1): Let $S = \{s_1, s_2, s_3, s_4\}$ and $\bar{S} = \{\bar{s}_1, \bar{s}_2, \bar{s}_3\}$, then r relates S to $\bar{S}$.

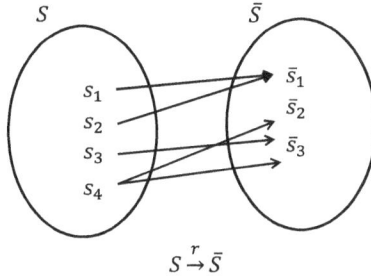

Figure 5.2-1

5.2.2 Statements (The Semantics of Realist Logic)

The form, "$r(s_i) = \bar{s}_j$," is called a "complete statement," which comprises part of realist logic's syntax. A statement can be complete or "incomplete." A statement is complete if r relates an element, $s_i \in S$, to an element, $\bar{s}_j \in \bar{S}$. Complete statements are either true or false. A statement is incomplete if r does not relate an element, $s_i \in S$, to an element, $\bar{s}_j \in \bar{S}$, or an element, $\bar{s}_j \in \bar{S}$, remains unrelated to an element, $s_i \in S$. Incomplete statements are neither true nor false.

If $s_i \in S$, $\bar{s}_i \in \bar{S}$, by fiat, $r(s_i) = \bar{s}_i$ is a true statement. Otherwise, it is false. The semantics of realist logic thus declare that complete statements are either true or false. For instance, as seen in Figure 5.2-1, the statement, "$r(s_1) = \bar{s}_1$," is true, whereas "$r(s_2) = \bar{s}_1$" is false.

5.2.3 Systems

A collection of statements, "$S \xrightarrow{r} \bar{S}$," where S contains n elements, "$s_i \in S, i = 0, \ldots, n$," or $\bar{S}$ contains m elements, "$\bar{s}_i \in \bar{S}, i = 0, \ldots, m$," is called a "system," $m, n \in \mathbb{N} \cup 0$, where $\mathbb{N}$ represents the set of counting numbers, i.e., $\mathbb{N} = \{1,2,3, \ldots\}$.

Either m or n, or indeed both, can be infinite. If the domain of S is $\mathbb{R}$, the real numbers, then the number of statements may not be countable. However, uncountable sets are not in the scope of this discussion.

5.2.3.1 Consistent and Inconsistent Systems

If $S \xrightarrow{r} \bar{S}$ is a system and

$$\exists s_i \exists s_j \big(s_i, s_j \in S \to r(s_i) = r(s_j) = \bar{s}_k \in \bar{S}, \qquad i \neq j \big),$$

then such statements are "confusions," meaning that more than one thought is associated with an instantiation of that thought — which is inconsistent.

Conversely, if $\bar{S} \xrightarrow{\bar{r}} S$ is a system such that

$$\exists \bar{s}_i \exists \bar{s}_j \big(\bar{s}_i, \bar{s}_j \in \bar{S} \to \bar{r}(\bar{s}_i) = \bar{r}(\bar{s}_j) = s_k \in S, \qquad i \neq j \big),$$

then such statements are "confoundings," which is also to say inconsistent. A confounding exists if more than one instantiation is associated with a thought. Confusions and confoundings are the only inconsistencies in realist logic.

A system is consistent if r is one-to-one. This is so because, if it were not, then, within an inconsistent system, either there exists

$$r(s_i) = r(s_j) = \bar{s}_k \in \bar{S}, \qquad i \neq j,$$

or

$$\bar{r}(\bar{s}_i) = \bar{r}(\bar{s}_j) = s_k \in S, \qquad i \neq j.$$

Either case or both is an inconsistency, and r is not one-to-one. Hence, a system, "$S \xrightarrow{r} \bar{S}$," is consistent if

$$\begin{aligned} \forall s_i \sim \exists s_j \big(s_i, s_j \in S \to r(s_i) = r(s_j) = \bar{s}_k \in \bar{S}, \qquad i \neq j \big) \\ \wedge \, \forall \bar{s}_i \sim \exists \bar{s}_j \big(\bar{s}_i, \bar{s}_j \in \bar{S} \to \bar{r}(\bar{s}_i) = \bar{r}(\bar{s}_j) = s_k \in S, \\ i \neq j \big). \end{aligned}$$

5.2.3.2 Complete and Incomplete Systems

The number of possible relationships, r, between S and $\bar{S}$ — where S contains n elements and $\bar{S}$ contains m elements — is 2^{nm}, since there are two sets, "$S, \bar{S}$," and the first set contains n elements and the second m. To see this, consider the ordered pair, $\langle s, \bar{s} \rangle$, where $s \in S$ is the first element and $\bar{s} \in \bar{S}$ is the second element of the ordered pair. Since each of the n elements in S can be related to each of the m elements in $\bar{S}$, by the fundamental principle of counting, the total possible number of ordered pairs that can be constructed between the sets, $S, \bar{S}$, is $n \cdot m$. Now a relationship, R, is defined as any collection of ordered pairs where the first element of the ordered pair belongs to S and the second to $\bar{S}$. Hence, for each ordered pair, $\langle s, \bar{s} \rangle$, either $\langle s, \bar{s} \rangle \in R$ or $\langle s, \bar{s} \rangle \notin R$. Since there are $n \cdot m$ ordered pairs, again, by the fundamental principle of counting, 2^{nm} relationships can be created since each ordered pair is either in a given relationship or absent from it.

However, an incomplete system is one in which, through r, an element in S is unrelated to an element in $\bar{S}$, or an element of $\bar{S}$ is unrelated to an element of S or both conditions might be the case.

The system, $S \xrightarrow{r} \bar{S}$, is complete if

$$\forall s_i \exists \bar{s}_j \big(s_i \in S \wedge \bar{s}_j \in \bar{S} \to r(s_i) = \bar{s}_j \big)$$
$$\wedge \exists s_i \forall \bar{s}_j \big(s_i \in S \wedge \bar{s}_j \in \bar{S} \to \bar{r}(\bar{s}_i) = s_j \big).$$

In other words, $S \xrightarrow{r} \bar{S}$ is a complete system if all its statements are complete. Note that a system can be complete but not consistent (Figure 5.2-1).

A system, $S \xrightarrow{r} \bar{S}$, has an "omission" if

$$\forall s_i \exists \bar{s}_j \big(s_i \in S \wedge \bar{s}_j \in \bar{S} \to r(s_i) \neq \bar{s}_j \big).$$

An omission is an instantiation with no accompanying thought.

A system, $S \xrightarrow{r} \bar{S}$, contains an "opinion" if

$$\exists s_i \forall \bar{s}_j \left(s_i \in S \wedge \bar{s}_j \in \bar{S} \rightarrow r(s_i) \neq \bar{s}_j \right).$$

An opinion is a thought with no instantiation. Omissions and opinions are the only incomplete statements within realist logic. A system is incomplete if it contains incomplete statements. Thus, a system is complete if and only if it has no incomplete statements. Incomplete statements are neither true nor false.

5.2.3.3 Equal and Equivalent Systems

A system's cardinality is designated by $\#S \xrightarrow{r} \bar{S}_{n,m}$, which signifies a nominal set with n elements and an instantiation set with m elements— $n, m \in \mathbb{N} \cup 0$. Two systems, $S \xrightarrow{r} \bar{S}$ and $T \xrightarrow{f} \bar{T}$, are syntactically equivalent if r and f have the same syntax. For instance (see Figure 5.2.3.3-1), $S \xrightarrow{r} \bar{S}$ and $T \xrightarrow{f} \bar{T}$ are syntactically, but not semantically, the same, designated $S \xrightarrow{r} \bar{S} \cong T \xrightarrow{f} \bar{T}$.

$$S \xrightarrow{r} \bar{S} \qquad\qquad\qquad T \xrightarrow{f} \bar{T}$$

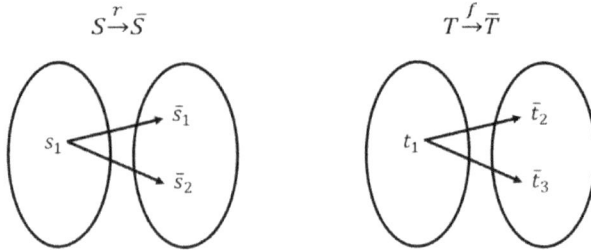

Figure 5.2.3.3-1

Two systems, $S \xrightarrow{r} \bar{S}$ and $T \xrightarrow{f} \bar{T}$, are syntactically equivalent if:

1. $\#S \xrightarrow{r} \bar{S} = \#T \xrightarrow{f} \bar{T}$.
2. $r \cong f$.

Two systems, $S \xrightarrow{r} \bar{S}$ and $T \xrightarrow{f} \bar{T}$, are equivalent if they have the same syntax and semantics, designated $S \xrightarrow{r} \bar{S} \equiv T \xrightarrow{f} \bar{T}$, as in:

1. $\#S \xrightarrow{r} \bar{S} = \#T \xrightarrow{f} \bar{T}$.
2. $r \equiv f$.

Two systems, $S \xrightarrow{r} \bar{S}$ and $T \xrightarrow{f} \bar{T}$, are equal if:

1. $\#S \xrightarrow{r} \bar{S} = \#T \xrightarrow{f} \bar{T}$.
2. $r \equiv f$.
3. $r = f$.

In other words, two systems are equal if they have the same cardinality, syntax and semantics, and statements $(r = f)$. Two systems are equivalent if they have the same cardinality, syntax and semantics.

5.2.3.4 System of Systems

Let

$$S = \left\{ S_i \xrightarrow{r_k} \bar{S}_j \right\}, \qquad \bar{S} = \left\{ \bar{S}_j \xrightarrow{\bar{f}_l} \bar{\bar{S}}_k \right\},$$

where $\left\{ S_i \xrightarrow{r_k} \bar{S}_j \right\}$ represent a collection of systems. Hence, $S_i \xrightarrow{r} \bar{\bar{S}}_k$ is a system of systems, where r now signifies a relationship between S_i and $\bar{\bar{S}}_k$. For example, the $ith - kth$ system might look like that shown in Figure 5.2.3.4-1.

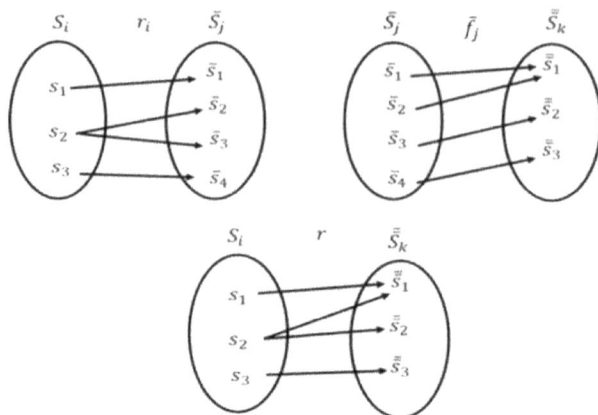

Figure 5.2.3.4-1

5.3 Complete and Consistent Systems

If a system, $S \xrightarrow{r} \bar{S}$, is consistent and complete, then r is one-to-one and onto. However, consistency and completeness are necessary, but not sufficient, conditions for system validity.

5.3.1 Valid Systems

Suppose the system, $S \xrightarrow{r} \bar{S}$, is complete and consistent, which is to say that r is one-to-one and onto. $S \xrightarrow{r} \bar{S}$ is valid if, and only if,

$$\forall s_i \forall \bar{s}_i (s_i \in S \wedge \bar{s}_i \in \bar{S} \leftrightarrow r(s_i) = \bar{s}_i).$$

In other words, the system is valid if and only if all its statements are true.

There is only one valid system within all possible complete and consistent systems, $S \xrightarrow{r} \bar{S}$. The necessary and sufficient conditions for validity are:

1. r is one-to-one and onto.
2. All the statements are true.

5.3.2 Complete and Consistent Invalid Systems

A complete and consistent system is entirely invalid if

$$\forall s_i \forall \bar{s}_i (s_i \in S \land \bar{s}_i \in \bar{S} \leftrightarrow r(s_i) \neq \bar{s}_i).$$

Put simply, the system is entirely invalid if and only if all its statements are false.

5.3.3 Complete and Consistent Partially Valid Systems

A complete and consistent system is partially valid if

$$\exists s_i \exists \bar{s}_i (s_i \in S \land \bar{s}_i \in \bar{S} \leftrightarrow r(s_i) = \bar{s}_i).$$

Note that a valid system is a partially valid system whose statements are all true.

5.3.4 Proposition on Complete and Consistent Systems

Proposition 1: If there are $n > 1 \in \mathbb{N}$ possible true statements in $S \xrightarrow{r} \bar{S}$ and

$$\#S \xrightarrow{r} \bar{S}_{n,n},$$

then there are $n!$ possible complete and consistent systems, $S \xrightarrow{r} \bar{S}$, one of which is valid, $(n-1)!$ of which are entirely invalid, and $n! - (n-1)! - 1$ are partially valid.

Proof: There are a total of $n!$ complete and consistent systems because there are n ways of selecting the first statement, $n - 1$ ways of selecting the second, $n - 2$ ways of selecting the third, etc. However, $n(n-1)(n-2)\ldots 1 = n!$. Since $\#S \xrightarrow{r} \bar{S}_{n,n}$, only one of the possible systems is valid because $r(s_i) = \bar{s}_i$ is a unique collection of statements. There are $n - 1$ choices for $r(s_1) \neq \bar{s}_1$, $n - 2$ choices for $r(s_1) \neq \bar{s}_1$, and $r(s_2) \neq \bar{s}_2$. Continuing, there are $(n-1)!$ possible entirely invalid systems. Of the remaining possibilities, some statements are true, and the rest of the statements are false. Therefore, the number of possible partially valid systems is $n! - (n-1)! - 1$. Hence, the theorem is proved.

Example: Let $\#S \xrightarrow{r} \bar{S}_{3,3}$ and let

$$S = \{s_1, s_2, s_3\}, \qquad \bar{S} = \{\bar{s}_1, \bar{s}_2, \bar{s}_3\}.$$

If so, there are $3! = 6$ possible complete and consistent systems, $S \xrightarrow{r} \bar{S}$. Note that the following system is valid:

$$r(s_1) = \bar{s}_1, \qquad r(s_2) = \bar{s}_2, \qquad r(s_3) = \bar{s}_3.$$

Since $(3-1)! = 2! = 2$, there are two possible entirely invalid systems:

$$r(s_1) = \bar{s}_2, \qquad r(s_2) = \bar{s}_3, \qquad r(s_3) = \bar{s}_1.$$

$$r(s_1) = \bar{s}_3, \qquad r(s_2) = \bar{s}_1, \qquad r(s_3) = \bar{s}_2.$$

Since $6 - 2 - 1 = 3$, there are three possible partially valid systems:

$$r(s_1) = \bar{s}_1, \qquad r(s_2) = \bar{s}_3, \qquad r(s_3) = \bar{s}_2.$$

$$r(s_2) = \bar{s}_2, \qquad r(s_1) = \bar{s}_3, \qquad r(s_3) = \bar{s}_1.$$

$$r(s_3) = \bar{s}_3, \qquad r(s_2) = \bar{s}_1, \qquad r(s_1) = \bar{s}_2.$$

5.4 Characteristics of a System

If $S \xrightarrow{r} \bar{S}$ is a system, let $T \subset S$ and $\bar{T} \subset \bar{S}$. $T \xrightarrow{f} \bar{T}$ would then be syntactically different from the original system if $f \neq r$. Two systems with no common statements are called "disjoint."

5.4.1 Subsystems

By insisting that $\bar{T}$ consists of all the elements of $\bar{S}$ that r, from T, associates with the elements in S, then

Definition: If $S \xrightarrow{r} \bar{S}$ is a system and $T \subset S$, then $T \xrightarrow{r} \bar{T}$ is called a "subsystem" of $S \xrightarrow{r} \bar{S}$.

If $T \xrightarrow{r} \bar{T}$ is a subsystem of $S \xrightarrow{r} \bar{S}$, this is designated:

$$T \xrightarrow{r} \bar{T} \subset S \xrightarrow{r} \bar{S}.$$

Example: In Figure 5.2-1, $S = \{s_1, s_2, s_3, s_4\}$ and $\bar{S} = \{\bar{s}_1, \bar{s}_2, \bar{s}_3\}$. Let $T = \{s_1, s_2\} \subset S$, then $\bar{T} = \{\bar{s}_1\} \subset \bar{S}$, since $r(s_1) = \bar{s}_1$ and $r(s_2) = \bar{s}_1$, which are the same statements as in $S \xrightarrow{r} \bar{S}$, but restricted to $T \xrightarrow{r} \bar{T}$. Hence,

$$T \xrightarrow{r} \bar{T} \subset S \xrightarrow{r} \bar{S}.$$

5.4.2 Universal Systems

Definition: A system, $U \xrightarrow{r} \bar{U}$, is termed universal if

$$\forall S \left(S \xrightarrow{r} \bar{S} \subset U \xrightarrow{r} \bar{U} \leftrightarrow S \subset U \right).$$

In other words, all the systems, $S \xrightarrow{r} \bar{S}$, are subsystems of $U \xrightarrow{r} \bar{U}$.

Proposition 2: If $U \xrightarrow{r} \bar{U}$ is a valid system and $S \xrightarrow{r} \bar{S}$ has at least one complete statement and is a subsystem of $U \xrightarrow{r} \bar{U}$, then $S \xrightarrow{r} \bar{S}$ is valid.

Proof: If $U \xrightarrow{r} \bar{U}$ is valid, then r is one-to-one and onto, and all of its statements are true. However, if $S \xrightarrow{r} \bar{S} \subset U \xrightarrow{r} \bar{U}$, then the statements in $S \xrightarrow{r} \bar{S}$ are identical to those in $U \xrightarrow{r} \bar{U}$, restricted to $S \xrightarrow{r} \bar{S}$. Hence, the statements in $S \xrightarrow{r} \bar{S}$ are one-to-one and onto, and true. Therefore, $S \xrightarrow{r} \bar{S}$ is valid, thereby completing the proof. In other words, all nonempty subsystems of valid systems are valid.

5.4.3 Propositions on Systems

Proposition 3: If $T \xrightarrow{r} \bar{T} \subset S \xrightarrow{r} \bar{S}$ and $T = S$, then $\bar{T} = \bar{S}$.

Proof: If $T \xrightarrow{r} \bar{T} \subset S \xrightarrow{r} \bar{S}$, then the statements in $T \xrightarrow{r} \bar{T}$ must be identical to those in $S \xrightarrow{r} \bar{S}$, restricted to $T \xrightarrow{r} \bar{T}$. However, if $T = S$, then $\bar{T} = \bar{S}$, since the statements in $T \xrightarrow{r} \bar{T}$ must be identical to those in $S \xrightarrow{r} \bar{S}$ but restricted to $T \xrightarrow{r} \bar{T}$, which can only happen if $\bar{T} = \bar{S}$. This, therefore, completes the proof.

Proposition 4: If $S \xrightarrow{r} \bar{S}$ is a system and

$$\exists s_i \exists! \bar{s}_i (s_i \in S \rightarrow r(s_i) = \bar{s}_i),$$

then

$$\exists T \exists \bar{T} \left(T \xrightarrow{r} \bar{T} \subset S \xrightarrow{r} \bar{S} \rightarrow T \xrightarrow{r} \bar{T} \text{ is a valid subsystem of } S \xrightarrow{r} \bar{S} \right).$$

Proof: Let $T \xrightarrow{r} \bar{T} \subset S \xrightarrow{r} \bar{S}$. If $T = \{s_i\} \subset S$ such that

$$\forall s_i \exists! \bar{s}_i (s_i \in T \rightarrow r(s_i) = \bar{s}_i \in \bar{T}),$$

then the statements belonging to $T \xrightarrow{r} \bar{T}$ are one-to-one and onto, and true. Hence, $T \xrightarrow{r} \bar{T}$ is valid, which completes the proof. Put simply, every system with at least one singular true statement has a valid subsystem.

Proposition 5: If $S \xrightarrow{r} \bar{S}$ is a system, then $\emptyset \xrightarrow{r} \bar{\emptyset}$ exists and is not incomplete.

Proof: Let $T \xrightarrow{r} \bar{T} \subset S \xrightarrow{r} \bar{S}$ and let $T = \emptyset \subset S$. If $\exists \bar{s}_i (\bar{s}_i \in \bar{T})$, then $T \xrightarrow{r} \bar{T}$ is an incomplete system since there are elements in $\bar{T}$, but none in T. However, $\bar{\emptyset} \subset \bar{S}$ by the ordinary rules of sets. Let $\bar{T} = \bar{\emptyset}$. Hence, there exists an entity, $\emptyset \xrightarrow{r} \bar{\emptyset}$, which is not incomplete since there are no incomplete statements in $\emptyset \xrightarrow{r} \bar{\emptyset}$, which proves the theorem.

On the other hand, $\emptyset \xrightarrow{r} \bar{\emptyset}$ is considered complete since there are no incomplete statements within it. Consequently, by fiat, $\emptyset \xrightarrow{r} \bar{\emptyset}$ is deemed a valid system since nothing (the empty set) is instantiated by "nothingness."

Proposition 6: If $S \xrightarrow{r} \bar{S}$ is a complete, consistent, and an entirely invalid system, then:

$$\exists T \exists \bar{T} \left(T \xrightarrow{r} \bar{T} \subset S \xrightarrow{r} \bar{S} \wedge T = \emptyset \wedge \bar{T} = \bar{\emptyset} \right.$$
$$\left. \to T \xrightarrow{r} \bar{T} \text{ is the only valid subsystem of } S \xrightarrow{r} \bar{S} \right).$$

Proof: Let $T \xrightarrow{r} \bar{T} \subset S \xrightarrow{r} \bar{S}$. Given that r is one-to-one and onto, and

$$\forall s_i (s_i \in S \to r(s_i) \neq \bar{s}_i),$$

then any nonempty subsystem, $T \xrightarrow{r} \bar{T}$, of $S \xrightarrow{r} \bar{S}$ is completely invalid.

Hence, $T = \emptyset \subset S \to \bar{T} = \bar{\emptyset} \subset \bar{S}$. Therefore, $T \xrightarrow{r} \bar{T}$ is valid. In other words, the only valid subsystem of a complete, consistent, and entirely invalid system is the empty subsystem, "$\emptyset \xrightarrow{r} \bar{\emptyset}$."

5.5 Set Operations on Valid Systems

Proposition 7: If $S \xrightarrow{r} \bar{S}$ and $R \xrightarrow{f} \bar{R}$ are valid systems, then $R \cap S \xrightarrow{y} \bar{R} \cap \bar{S}$ and $R \cup S \xrightarrow{g} \bar{R} \cup \bar{S}$ are valid.

Proof: There are two cases for each claim: 1) $R \cap S = \emptyset$ and 2) $R \cap S \neq \emptyset$. If $R \cap S = \emptyset$, assume $\bar{R} \cap \bar{S} \neq \bar{\emptyset}$. Hence, there would exist an element, $\bar{s}_i \in \bar{R} \cap \bar{S}$, such that $g(s_i) = \bar{s}_i$, $s_i \in S, R$ since both $S \xrightarrow{r} \bar{S}$ and

$R \xrightarrow{f} \bar{R}$ are valid. But this contradicts the assumption that $R \cap S = \emptyset$. Hence,

$$R \cap S \xrightarrow{g} \bar{R} \cap \bar{S} \rightarrow \emptyset \xrightarrow{g} \bar{\emptyset},$$

which is valid. If $R \cap S \neq \emptyset$, then $\exists s(s \in S, s \in R)$. And yet, $r(s) = \bar{s}$ and $f(s) = \bar{s}$, since both $S \xrightarrow{r} \bar{S}$ and $R \xrightarrow{f} \bar{R}$ are valid. Hence, $\bar{s} \in \bar{R} \cap \bar{S}$. Moreover, there exists a function, g, such that for all s, $g(s) = \bar{s}$. Since s is arbitrary, $R \cap S \xrightarrow{g} \bar{R} \cap \bar{S}$ is valid.

If $R \cap S = \emptyset$, then

$$\forall i \forall j (r_i \neq s_j, \qquad r(s_i) = \bar{s}_i, \qquad f(r_i) = \bar{r}_i),$$

since both $S \xrightarrow{r} \bar{S}$ and $R \xrightarrow{f} \bar{R}$ are valid. Hence,

$$R \cup S \xrightarrow{g} \bar{R} \cup \bar{S} \rightarrow \forall s_i \forall r_i \exists g (s_i \in S \wedge r_i \in R \rightarrow g(s_i) = \bar{s}_i, \qquad g(r_i) = \bar{r}_i, \\ s_i \neq r_j).$$

Therefore, $R \cup S \xrightarrow{g} \bar{R} \cup \bar{S}$ is valid. If $R \cap S \neq \emptyset$, then $\exists s(s \in S, s \in R)$ where s is a common element between R and S. In any case, there exists a g such that for all s, $g(s) = \bar{s}$. Since s is arbitrary, $R \cup S \xrightarrow{g} \bar{R} \cup \bar{S}$ is valid, which completes the proof.

Example: Let

$$R = \{a, b\}, \qquad \bar{R} = \{\bar{a}, \bar{b}\}, \qquad S = \{a, c, d\}, \qquad \bar{S} = \{\bar{a}, \bar{c}, \bar{d}\}.$$

If $R \xrightarrow{r} \bar{R}$ and $S \xrightarrow{f} \bar{S}$ are valid systems, then

$$r(a) = \bar{a}, \qquad r(b) = \bar{b}, \qquad f(a) = \bar{a}, \qquad f(c) = \bar{c}, \qquad f(d) = \bar{d}.$$

Now $R \cap S = \{a\}$ and $\bar{R} \cap \bar{S} = \{\bar{a}\}$. But $r(a) = \bar{a}$ and $f(a) = \bar{a}$. Hence, $R \cap S \xrightarrow{g} \bar{R} \cap \bar{S}$ is valid, since $g(a) = \bar{a}$. Now

$$R \cup S = \{a, b, c, d\} \text{ and } \bar{R} \,\bar{\cup}\, \bar{S} = \{\bar{a}, \bar{b}, \bar{c}, \bar{d}\}.$$

As well as:

$$r(a) = \bar{a}, \qquad r(b) = \bar{b}, \qquad f(a) = \bar{a}, \qquad f(c) = \bar{c}, \qquad f(d) = \bar{d}.$$

Hence, $R \cup S \xrightarrow{g} \bar{R} \,\bar{\cup}\, \bar{S}$ is valid, since g exists such that $g(a) = \bar{a}$, and so on.

5.6 Functions on Systems

Proposition 8: Suppose the system, $A \xrightarrow{r} \bar{A}$, is valid and let $B \xrightarrow{f} \bar{B}$ be a system. If a function, g, exists such that $A \xrightarrow{g} B$, as well as a function, $\bar{g}$, such that

$$\forall a \big(a \in A \wedge g(a) = b \in B \rightarrow \bar{g}(\bar{a}) = \bar{b} \in \bar{B} \big),$$

then $\left(B \xrightarrow{g_c} A \right) \xrightarrow{t} \left(\bar{A} \xrightarrow{\bar{g}} \bar{B} \right)$ is a valid system.

Proof: Suppose $g(a) = b$, $a \in A$, $b \in B$, then $g_c(b) = a$, where g_c is the converse of g. Since $A \xrightarrow{r} \bar{A}$ is valid, $\forall a(a \in A \rightarrow r(a) = \bar{a})$, and, furthermore, $\bar{g}(\bar{a}) = \bar{b}$. Therefore, there exists a function, t, such that $\forall b \big(t(b) = \bar{b} \big)$. Since b is arbitrary, $\left(B \xrightarrow{g_c} A \right) \xrightarrow{t} \bar{A} \xrightarrow{\bar{g}} \bar{B}$ is valid, thus proving the proposition. Note that $\bar{g}$ instantiates g.

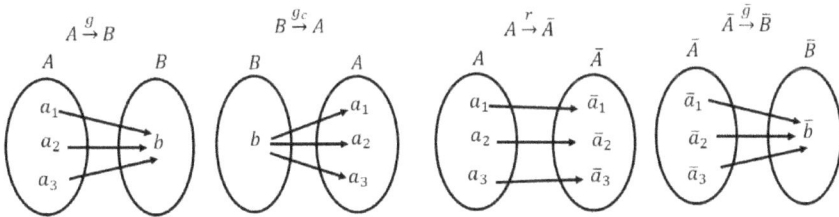

Figure 5.6-1

Example: In the above figure, the function $A \overset{g}{\to} B$ sends all the a_i's into b. Hence, $B \overset{g_c}{\to} A$ sends b into the a_i's. Since $A \overset{r}{\to} \bar{A}$ is valid,

$$\forall a_i (a_i \in A \to r(a_i) = \bar{a}_i).$$

Hence, r sends all the a_i's into the $\bar{a}_i$'s. Finally, $\bar{A} \overset{\bar{g}}{\to} \bar{B}$ sends $\bar{a}_i$'s into $\bar{b}$. Therefore, there is a function, t, that sends b into $\bar{b}$, which is valid.

5.6.1 Composite Functions on Systems

Proposition 9: Suppose the system, $A \overset{r}{\to} \bar{A}$, is valid, and $B \overset{s}{\to} \bar{B}$ and $C \overset{t}{\to} \bar{C}$ are systems. Let $\left(A \overset{f}{\to} B \overset{g}{\to} C\right)$, where f and g are functions. If

$$f(a) = b \wedge g(b) = c \to \bar{f}(\bar{a}) = \bar{b} \wedge \bar{g}(\bar{b}) = \bar{c}, \ a \in A, \ b \in B, \ c \in C,$$

then $\left(C \overset{(g \circ f)_c}{\longrightarrow} A\right) \overset{h}{\to} \left(\bar{A} \overset{\bar{g} \circ \bar{f}}{\longrightarrow} \bar{C}\right)$ is valid.

Proof: $f(a) = b \wedge g(b) = c \to \bar{f}(\bar{a}) = \bar{b} \wedge \bar{g}(\bar{b}) = \bar{c}$, then

$$(g \circ f)(a) = c \to (g \circ f)_c(c) = a.$$

Since $A \overset{r}{\to} \bar{A}$ is valid, $\forall a(r(a) = \bar{a})$ and $(\bar{g} \circ \bar{f})(\bar{a}) = \bar{c}$. Hence, a function, h, exists such that $h(c) = \bar{c}$. Since c is arbitrary,

$$\left(C \overset{(g \circ f)_c}{\longrightarrow} A\right) \overset{h}{\to} \left(\bar{A} \overset{\bar{g} \circ \bar{f}}{\longrightarrow} \bar{C}\right)$$

is valid, which proves the theorem.

5.6.2 Valid Functions

Proposition 10: Suppose the system, $A \overset{r}{\to} \bar{A}$, is valid and $B \overset{s}{\to} \bar{B}$ is a system. If $\{f, g, h, ...\}$ and $\{\bar{f}, \bar{g}, \bar{h}, ...\}$ are sets of functions from A into B and from $\bar{A}$ into $\bar{B}$, respectively, then

$$\{f,g,h,...\} \overset{t}{\to} \{\bar{f},\bar{g},\bar{h},...\}$$

is valid if and only if $t(f) = \bar{f}$, $t(g) = \bar{g}$, $t(h) = \bar{h}$, ..., where

$$\forall a \forall b \forall \bar{f}\big(a \in A, b \in B \wedge f(a) = b \to \bar{f}(\bar{a}) = \bar{b}\big).$$

Proof: Since $A \overset{r}{\to} \bar{A}$ is valid,

$$\forall a(a \in A \to r(a) = \bar{a}).$$

If $f(a) = b$, then $f_c(b) = a$ and $r(a) = \bar{a}$. For $\left(B \overset{f_c}{\to} A\right) \overset{t}{\to} \left(\bar{A} \overset{\bar{f}}{\to} \bar{B}\right)$ to be valid, it must be the case that $\bar{f}(\bar{a}) = \bar{b}$ since $f(a) = b$. In only this case can a function, t, exist such that

$$\forall b\big(t(b) = \bar{b}\big).$$

Moreover, if $f \overset{t}{\to} \bar{f}$, then $f(a) = b$ implies that $\bar{f}(\bar{a}) = \bar{b}$. But then t sends b into $\bar{b}$. Since a, b, and f are arbitrary,

$$\forall f \exists t \left(f \overset{t}{\to} \bar{f} \to \left(B \overset{f_c}{\to} A\right) \overset{t}{\to} \left(\bar{A} \overset{\bar{f}}{\to} \bar{B}\right) \text{ is valid}\right),$$

which completes the proof.

Moreover, if $A \overset{r}{\to} \bar{A}$ is valid, then r is one-to-one and onto, and hence, r has an inverse function, $\bar{r}$, such that $r(a) = \bar{a} \leftrightarrow \bar{r}(\bar{a}) = a$. Hence,

$$(\bar{r} \circ r)(a) = a, \qquad (r \circ \bar{r})(\bar{a}) = \bar{a}.$$

Therefore,

$$\forall a \forall \bar{a} \exists f \left((\bar{r} \circ r)(a) \overset{f}{\to} (r \circ \bar{r})(\bar{a})\right)$$

is valid, where f sends a into $\bar{a}$ for all a.

5.7 Relations

If $A \xrightarrow{r} \bar{A}$ and $B \xrightarrow{f} \bar{B}$ are systems and there exists a function, g:

$$A \times B \xrightarrow{g} \bar{A} \times \bar{B},$$

where g sends a group of ordered pairs,

$$\langle x, y \rangle \in A \times B, \qquad x \in A, \qquad y \in B,$$

into another group of ordered pairs,

$$\langle \bar{a}, \bar{b} \rangle \in \bar{A} \times \bar{B}, \qquad \bar{a} \in \bar{A}, \qquad \bar{b} \in \bar{B},$$

then $A \times B \xrightarrow{g} \bar{A} \times \bar{B}$ is a system of relations.

The ordered pair, $\langle x, y \rangle$, is called a "relation" in A and B. If

$$\exists x \exists y \exists g (\langle x, y \rangle \in A \times B \rightarrow g[\langle x, y \rangle] = \langle \bar{x}, \bar{y} \rangle \in \bar{A} \times \bar{B}),$$

then, by fiat, the relation is true. Otherwise, it is false.

5.7.1 A Valid System of Relations

Given a system of relations, $A \times B \xrightarrow{g} \bar{A} \times \bar{B}$, let $C' \subset A \times B$ such that

$$\forall a \forall b \forall \bar{a} \forall \bar{b} \exists g (\langle a, b \rangle \in C' \subset A \times B \wedge \langle \bar{a}, \bar{b} \rangle \in \bar{C}' \subset \bar{A} \times \bar{B} \rightarrow g(\langle a, b \rangle)$$
$$= \langle \bar{a}, \bar{b} \rangle \in \bar{C}'),$$

then $C' \xrightarrow{g} \bar{C}'$ is a valid system of relations. In other words, all the relations belonging to C' are true, and g is one-to-one and onto.

For example, if

$$\forall a (\langle a, a \rangle \in C' \subset A \times A \rightarrow g(\langle a, a \rangle) = \langle \bar{a}, \bar{a} \rangle \in \bar{C}'),$$

then the relationship is reflexive. If

$$\forall a \exists! \, b\big((a, b) \in C' \subset A \times B \rightarrow g((a, b)) = (\bar{a}, \bar{b}) \in \bar{C}'\big),$$

then the relation is a function.

5.8 The Laws of Valid Systems

Table 5.8-1 depicts the algebra of valid systems. If $U \xrightarrow{r} \bar{U}$ is valid, then so are all of its subsystems. The laws of valid systems are similar to those of set operations associated with Cantorian set theory.

The Algebra of Valid Systems	
$U \xrightarrow{r} \bar{U}, \quad A, B, C \subset U$	
Idempotent Laws	
1a. $A \cup A \xrightarrow{r} \bar{A} \bar{\cup} \bar{A}$	1b. $A \cap A \xrightarrow{r} \bar{A} \bar{\cap} \bar{A}$
Associative Laws	
2a. $(A \cup B) \cup C = A \cup (B \cup$ $C) \xrightarrow{r} (\bar{A} \bar{\cup} \bar{B}) \bar{\cup} \bar{C} = \bar{A} \bar{\cup} (\bar{B} \bar{\cup} \bar{C})$	2b. $(A \cap B) \cap C = A \cap (B \cap$ $C) \xrightarrow{r} (\bar{A} \bar{\cap} \bar{B}) \bar{\cap} \bar{C} = \bar{A} \bar{\cap} (\bar{B} \bar{\cap} \bar{C})$
Commutative Laws	
3a. $A \cup B = B \cup A \xrightarrow{r} \bar{A} \bar{\cup} \bar{B} =$ $\bar{B} \bar{\cup} \bar{A}$	3b. $A \cap B = B \cap A \xrightarrow{r} \bar{A} \bar{\cap} \bar{B} = \bar{B} \bar{\cap} \bar{A}$
Distributive Laws	
4a. $(A \cup (B \cap C) = (A \cup B) \cap (A \cup$ $C) \xrightarrow{r} (\bar{A} \bar{\cup} (\bar{B} \bar{\cap} \bar{C}) =$ $(\bar{A} \bar{\cup} \bar{B}) \bar{\cap} (\bar{A} \bar{\cup} \bar{C})$	4b. $(A \cap (B \cup C) = (A \cap B) \cup (A \cap$ $C) \xrightarrow{r} (\bar{A} \bar{\cap} (\bar{B} \bar{\cup} \bar{C}) =$ $(\bar{A} \bar{\cap} \bar{B}) \bar{\cup} (\bar{A} \bar{\cap} \bar{C})$
Identity Laws	
5a. $A \cup \emptyset = A \xrightarrow{r} \bar{A} = \bar{A} \bar{\cup} \bar{\emptyset}$	5b. $A \cap U = A \xrightarrow{r} \bar{A} = \bar{A} \bar{\cap} \bar{U}$
6a. $A \cup U = U \xrightarrow{r} \bar{U} = \bar{A} \bar{\cup} \bar{U}$	6b. $A \cap \emptyset = \emptyset \xrightarrow{r} \bar{\emptyset} = \bar{A} \bar{\cap} \bar{\emptyset}$
Compliment Laws	
7a. $A \cup \sim A = U \xrightarrow{r} \bar{U} = \bar{A} \bar{\cup} \sim\bar{A}$	7b. $A \cap \sim A = \emptyset \xrightarrow{r} \bar{\emptyset} = \bar{A} \bar{\cap} \sim\bar{A}$
8a. $\sim\sim A = A \xrightarrow{r} \bar{A} = \sim\sim\bar{A}$	8b. $\sim U = \emptyset \xrightarrow{r} \bar{\emptyset} = \sim\bar{U}, \sim\emptyset =$ $U \xrightarrow{r} \bar{U} = \sim\bar{\emptyset}$
De Morgan's Laws	
9a $\sim(A \cup B) = \sim A \cap$ $\sim B \xrightarrow{r} \sim(\bar{A} \bar{\cup} \bar{B}) = \sim\bar{A} \bar{\cap} \sim\bar{B}$	9b. $\sim(A \cap B) = \sim A \cup$ $\sim B \xrightarrow{r} \sim(\bar{A} \bar{\cap} \bar{B}) = \sim\bar{A} \bar{\cup} \sim\bar{B}$

Table 5.8-1

The following example illustrates the importance of valid systems: Suppose $A \xrightarrow{r} \bar{A}$ is valid, but $B \xrightarrow{s} \bar{B}$ is complete and consistent yet entirely invalid. If $A \cap B = \emptyset$ but $\bar{A} \bar{\cap} \bar{B} \neq \bar{\emptyset}$, then

$$A \cap B \xrightarrow{t} \bar{A} \bar{\cap} \bar{B}$$

is an incomplete system. If $\bar{A} \bar{\cap} \bar{B} = \bar{\emptyset}$, then

79

$$A \cap B \overset{t}{\to} \bar{A} \cap \bar{B}$$

is valid. And yet $A \cup B \overset{t}{\to} \bar{A} \cup \bar{B}$ is partially valid, since

$$\exists a \exists b \big(a \in A, b \in B \to r(a) = \bar{a} \wedge s(b) \neq \bar{b}\big).$$

Now suppose $A \cap B \neq \emptyset$, and that $a, b \in A, B$ implies

$$r(a) = \bar{a}, \qquad r(b) = \bar{b}, \qquad s(a) = \bar{c}, \qquad s(b) = \bar{d}, \qquad \bar{a} \neq \bar{c},$$
$$\bar{b} \neq \bar{d}.$$

Therefore,

$$A \cap B = \{a, b\} \cap \{a, b\} = \{a, b\} \overset{t}{\to} \{\bar{a}, \bar{b}\} \cap \{\bar{c}, \bar{d}\} = \bar{\emptyset},$$

which is incomplete. On the other hand,

$$A \cup B = \{a, b\} \cup \{a, b\} = \{a, b\} \overset{t}{\to} \{\bar{a}, \bar{b}\} \cup \{\bar{c}, \bar{d}\} = \{\bar{a}, \bar{b}, \bar{c}, \bar{d}\},$$

which is either invalid or incomplete.

The operations shown in Table 5.8-1 are valid if performed on valid systems. Simply put, set operations performed on valid systems are valid.

5.8.1 Valid Operators

Consider a set of operators, $O = \{\cup, \cap, \Delta, \dots\}$, and the set of operator instantiations, $\bar{O} = \{\bar{\cup}, \bar{\cap}, \bar{\Delta}, \dots\}$. The system, $O \overset{r}{\to} \bar{O}$, consists of valid operators if and only if

$$r(\cup) = \bar{\cup}, \qquad r(\cap) = \bar{\cap}, \qquad r(\Delta) = \bar{\Delta}, \dots$$

To see this, binary operators are functions, as in:

$$\forall x \forall y \exists z \exists f \left(x, y, z \in A \rightarrow (x \circ y) \overset{f}{\rightarrow} z \in A\right),$$

where "$\circ$" is a binary operator. A system of binary operators, $0 \overset{r}{\rightarrow} \bar{0}$, is valid if and only if

$$\left((x \circ y) \overset{f}{\rightarrow} z\right) \overset{r}{\rightarrow} \left((\bar{x} \bar{\circ} \bar{y}) \overset{\bar{f}}{\rightarrow} \bar{z}\right), \qquad f(\circ) = \bar{\circ} \in \bar{0}, \qquad \circ \in 0.$$

In other words, $\bar{f}$ must be a valid instantiation of f.

5.9 Paradoxes

Realist logic does not permit a valid system of all systems. A formal proof follows: Let $S \overset{f}{\rightarrow} \bar{S}$ be a system of systems, where $S = \{S_1, S_2, ...\}$ is the system of all systems. According to realist logic, S must have an instantiation set, $\bar{S} = \{\bar{S}_1, \bar{S}_2, ...\}$. However, a system of all systems requires that $\bar{S} \in S$, i.e.,

$$S = \{S_1, S_2, ..., \bar{S}_1, \bar{S}_2, ...\}.$$

Yet each of the S_i's must be instantiated by $\bar{S}_i$'s, i.e.,

$$\bar{S}' = \{\bar{S}_1, \bar{S}_2, ..., \bar{\bar{S}}_1, \bar{\bar{S}}_2, ...\}.$$

But then, S and $\bar{S}'$ would share common elements, thereby violating the definition of a system. This completes the proof.

Example: The barber paradox. All the men in a town are shaved either by themselves or by the barber. Let A be the set of all men in the town and let B be the set, "shave themselves" (s) or "shaved by the barber" (b), then

$$x, y, ... \in A, \qquad s, b \in B.$$

Let $S \subset A$ such that

$$\forall x(x \in S \to \langle x, s \rangle \in C \subset A \times B).$$

Hence, C contains those men who shave themselves. Let $T \subset A$ such that

$$\forall x(x \in T \to \langle x, b \rangle \in C' \subset A \times B).$$

Hence, the set, C', contains those men shaved by the barber. The assumption is that:

$$S \cap T = \emptyset.$$

Let $M \subset A = \{a, b, c\}$ signify three men who live in the town, where b designates the barber. Let $\langle a, s \rangle, \langle b, s \rangle \in C$ and $\langle c, b \rangle, \langle b, b \rangle \in C'$. Hence, let $D = \{\langle a, s \rangle, \langle b, s \rangle\}$ and $E = \{\langle c, b \rangle, \langle b, b \rangle\}$, then let

$$R' = D \cup E = \{\langle a, s \rangle, \langle b, s \rangle, \langle c, b \rangle, \langle b, b \rangle\}.$$

The possible instantiations consist of the following set of relations:

$$\bar{R} = \{\langle \bar{a}, \bar{s} \rangle, \langle \bar{a}, \bar{b} \rangle, \langle \bar{b}, \bar{s} \rangle, \langle \bar{c}, \bar{s} \rangle, \langle \bar{c}, \bar{b} \rangle\},$$

where $\langle \bar{a}, \bar{s} \rangle$ signifies that a shaves himself and $\langle \bar{a}, \bar{b} \rangle$ signifies that b shaves a and so on. Therefore,

$$\bar{R}' = \{\langle \bar{a}, \bar{s} \rangle, \langle \bar{b}, \bar{s} \rangle, \langle \bar{c}, \bar{b} \rangle\},$$

where $\bar{R}'$ contains only the set of true relations (assuming the barber is shaved). However, as the system, $R' \overset{r}{\to} \bar{R}'$, is either inconsistent or incomplete, since $\#\bar{R}' \neq \#R'$, it cannot be valid (Figure 5.9-1).

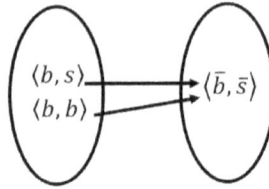

Figure 5.9-1

By creating a two-set system of logic, the paradoxes associated with Cantorian set theory disappear.

5.10 Concluding Remarks

Realist logic affirms the existence of universal and abstract mathematical objects. Defined in terms of two sets, the logic has a nominal set, S, representing an undivided entity, and a set, $\bar{S}$, which represents instantiated undividedness. The primary notion within realist logic is the "system," denoted $S \overset{r}{\to} \bar{S}$, where r defines a relationship between the elements in S and $\bar{S}$.

A system, $S \overset{r}{\to} \bar{S}$, is complete if

$$\forall s_i \exists \bar{s}_j \left(s_i \in S \wedge \bar{s}_j \in \bar{S} \to r(s_i) = \bar{s}_j \right)$$
$$\wedge \exists s_i \forall \bar{s}_j \left(s_i \in S \wedge \bar{s}_j \in \bar{S} \to \bar{r}(\bar{s}_i) = s_j \right).$$

Otherwise, it is incomplete.

A system, $S \overset{r}{\to} \bar{S}$, is consistent if r is one-to-one.

If $S \overset{r}{\to} \bar{S}$ is a complete and consistent system such that

$$\forall s \forall \bar{s} (s \in S \wedge \bar{s} \in \bar{S} \to r(s) = \bar{s}),$$

then $S \overset{r}{\to} \bar{S}$ is a valid system, where $r(s) = \bar{s}$ is a complete statement. Moreover, if $s \in S$ and $r(s) = \bar{s}$, then the statement is true. If not, it is false. A relationship, r, that is onto, but not a function, is a confusion, where at least one distinct thought is associated with more than one instantiation. A relationship, $\bar{r}$, that is onto, but not a function, represents a confounding, where at least one distinct instantiation is associated with more than one thought. If r is not onto, the system is incomplete. An incomplete system contains incomplete statements. If $\bar{r}$ is not onto, then an element $s \in S$ is an opinion in that it has no instantiation.

All valid systems are consistent, complete, and all their statements are true. All subsystems of valid systems are valid. Set operations on systems are valid if and only if the operations are performed on valid systems.

The set, C, is a relation in A and B if $\langle a, b \rangle \in C$, $a \in A$, $b \in B$. If

$$\langle \bar{a}, \bar{b} \rangle \in \bar{A} \times \bar{B} \rightarrow t(\langle a, b \rangle) = \langle \bar{a}, \bar{b} \rangle,$$

the relation, $\langle a, b \rangle$, is true. Otherwise, it is false.

Finally, developing a two-set logic model can help avoid the paradoxes associated with Cantorian set theory. While other theorems associated with realist logic are not explored here, the examples and discussion provide a sufficient introduction to the subject.

Epilogue

What's the greatest discovery in the history of thought? Of course, it's a silly question – but it won't stop me from suggesting an answer. It's Plato's discovery of abstract objects.

– James Robert Brown, Philosophy of Mathematics

This book, written from the perspective of "logicism," centers around the belief that all mathematics reduces to logic. Russell was logicism's foremost advocate. The *Principia Mathematica* and his subsequent writings were attempts at validating this belief. His validation almost succeeded, but not quite. As a foundation for mathematics, his "theory of types" required additional ad hoc strengthening beyond the fundamentals of logic. Nevertheless, set theory, widely regarded as the definitive work in the foundations of mathematics, turned out to be practically equivalent to first-order logic and gave credence to the opinion that logic and mathematics belonged to the same branch of learning.

Notably, logicism is not the only approach to answering the question, "where do numbers come from?" Intuitionism and formalism comprise the other schools of thought on this question [291]. Each of the schools has its strong as well as its weak points [292]. But logicism is, by far, the most common view, which suggests that the question of "what is mathematics?" is far from settled.

Following the rise of logicism, questions arose concerning the nature of logic. Logicians generally agreed on the syntactical rules that encompassed traditional propositional logic. The debates mostly centered around the semantics of the subject. Were the rules of logic merely inventions, or were they actual discoveries? Nominalistic philosophers primarily argued that mathematics was a type of formal language, hence, an invention of sorts.

On the other hand, realist philosophers argued that mathematical objects were timeless entities independent of language, thought, and

85

practices. Mathematics was, therefore, a series of discoveries. This book attempts to broaden the debate by showing that if the realist position is taken seriously, not only do semantical issues arise, but the syntactical rules that emerge are markedly different from those associated with traditional propositional logic.

If, indeed, each logic produces a different set of syntactical rules, then it is reasonable to ask, "which logic system is the correct one?" Not surprisingly, this question has no simple answer. Beyond their respective starting points, each approach appears meticulously developed. Barring outright miscalculation or an overlooked omission, both approaches appear equally valid. If so, this gets no closer to answering the question, "which logic system is the correct one?" The nominalist view would no doubt be that the two logics are simply different languages, different inventions.

Conversely, a realist would maintain the instantiation set associated with realist logic is independent of thought, language, and practices, and hence, exists. And if it exists, it must be included in any logic system, suggesting that traditional logic suffers from missing pieces.

On the other hand, possibly, the two approaches only appear to have a separate set of syntactical rules but, upon closer scrutiny, would not. For instance, the modern idea of a set is expressed, "$\{x:\theta\}$," where the $x's$ are constrained by the property, θ. Hence, it requires two notions to define a "set." The $x's$ are the substance of the set, while θ defines the attributes of the $x's$. But once this is accomplished, $\{x:\theta\}$ becomes a unit of sorts. This approach appears eerily similar to the realist approach, where two independent sets are required to define the system. However, the primitive idea in realist logic is the "relation," $\langle S,\bar{S}\rangle$, given in traditional set notation as $\langle S,\bar{S}\rangle = \langle\{x:\theta\},\{\bar{x}:\bar{\theta}\}\rangle$. In opposition, traditional propositional logic creates relations from more fundamental concepts, namely a subset of the cross-product of two sets. In fact, a relation could be comprised of elements from the same set. Hence, the two approaches seem fundamentally different, at least in that respect. Moreover, realist logic contains a universal system that traditional set theory forbids, which, if included, creates a contradiction. Realist logic appears to avoid a contradictory "system of all systems" by forbidding singular notions of this type.

Moreover, the null or empty set is defined in terms of a universal set and vice versa, which is the source of the conflict. Thus, if any reference to a universal set is purged altogether, the inconsistency vanishes since a universal set is eliminated from the logic system. Such is the method most commonly pursued by logicians when creating a theory of sets. By contrast, realist logic's universal system is defined by a separate "universalness" set, avoiding reference to the null system. This way, no conflict emerges between the null system and the universal system. Whether this characteristic of realist logic represents an advantage over traditional logic remains a question for philosophers.

A better question might be, "which of the two logics is the most useful?" The usefulness of traditional logic can hardly be debated. Its utility in human endeavors such as trade, finance, engineering, and many other activities necessary for human comfort and the well-being of humans can scarcely be questioned. But, of course, mathematics existed long before the advent of set theory or predicate calculus. Nevertheless, putting mathematics on such a solid foundation gave Mankind confidence that the methods employed by its users were indeed reliable. Realist logic, on the other hand, is newer, so its utility is less obvious. However, suppose indeed that the syntactical rules of realist logic are genuinely different from traditional logic. In that case, it is safe to assume that the number systems that emerge will be somewhat, if not significantly, different than those that arise from traditional logic. The extent to which these two logics produce divergent number systems is a matter for further study.

Why might this divergence be significant? First, modern physics is a product of two worlds. Both worlds rely on mathematics to convey various predictions about the nature of "our universe." The theory of relativity predicts what happens in the large-scale universe, regulated mainly by gravity. While sophisticated and incredibly accurate in its calculations, the theory uses only ordinary numbers to make its predictions. And ordinary numbers spring from traditional logic.

Second, the theory of relativity is primarily a study in geometry, which also has a solid foundation in traditional logic.

Quantum mechanics, which makes predictions about the small-scale physics of "our universe," tells a different story. While quantum calculations are incredibly accurate, the lack of an agreed-upon ontology regarding the nature of quantum mechanics is a source of embarrassment. Even the most ardent defenders of the subject admit that it is intellectually opaque. Some physicists are blunter, arguing that "quantum mechanics doesn't make sense" [14]. It is probably best for physicists to come clean and admit that, at present, no one knows what the equations of quantum mechanics purportedly say. But the answer typically provided is somewhat different. It's not that physicists don't know the answer; they cannot agree on the answer. At any rate, the mathematical foundations that support quantum mechanics can not be derived from traditional propositional logic. More demonstratively, the truth of this claim is embodied in a mathematical theorem known as Bell's Theorem.

The situation is of great frustration to physicists, who would like to find just one theory that explains the nature of the universe. Moreover, the quest to unify the theory of relativity with quantum mechanics is exacerbated because the two theories are logically incompatible. Might a different approach to logic provide a vehicle that avoids the roadblocks that currently prevent physicists from offering a more comprehensive theory? Preliminary research has shown that realist logic shows the promise of an entirely new and helpful direction. However, a better answer to such a question requires additional study.

Appendix

If $S(x)$ and $T(x)$ are formulas and x is not free, then the following are logically equivalent [178]:

Logically Equivalent Sentences with Quantifiers
1. $\forall(x)S(x) \equiv {\sim}\exists(x){\sim}S(x)$
2. $\forall(x){\sim}S(x) \equiv {\sim}\exists(x)S(x)$
3. $\exists(x)S(x) \equiv {\sim}\forall(x){\sim}S(x)$
4. $\exists(x){\sim}S(x) \equiv {\sim}\forall(x)S(x)$
5. $\forall(x)\forall(y)S(x,y) \equiv \forall(y)\forall(x)S(x,y)$
6. $\exists(x)\exists(y)S(x,y) \equiv \exists(y)\exists(x)S(x,y)$
7. $\exists(x)\forall(y)S(x,y) \equiv \forall(y)\exists(x)S(x,y)$
8. $\forall(x)(S(x) \wedge T(x)) \equiv \forall(x)S(x) \wedge \forall(x)T(x)$
9. $\exists(x)(S(x) \vee T(x)) \equiv \exists(x)S(x) \vee \exists(x)T(x)$
10. $\forall(x)(S(x) \rightarrow T(x)) \equiv \exists(x)S(x) \rightarrow \exists(x)T(x)$
11. $\exists(x)(S(x) \wedge T(x)) \equiv \exists(x)S(x) \wedge \exists(x)T(x)$
12. $\forall(x)S(x) \vee \forall(x)T(x) \equiv \forall(x)(S(x) \vee T(x))$
13. $\forall(x)(S(x) \vee T(x)) \equiv S(x) \vee \forall(x)T(x)$
14. $\exists(x)(S(x) \vee T(x)) \equiv S(x) \vee \exists(x)T(x)$
15. $\forall(x)(S(x) \wedge T(x)) \equiv S(x) \wedge \forall(x)T(x)$
16. $\exists(x)(S(x) \wedge T(x)) \equiv S(x) \wedge \exists(x)T(x)$
17. $\forall(x)(S(x) \rightarrow T(x)) \equiv S(x) \rightarrow \forall(x)T(x)$
18. $\exists(x)(S(x) \rightarrow T(x)) \equiv S(x) \rightarrow \exists(x)T(x)$
19. $\forall(x)(S(x) \rightarrow T(x)) \equiv \exists(x)S(x) \rightarrow T(x)$
20. $\exists(x)(S(x) \rightarrow T(x)) \equiv \forall(x)S(x) \rightarrow T(x)$

The table shows that in a proof, a sentence on the left of "$\equiv$" may be substituted for a sentence on the right of "$\equiv$" and vice versa.

List of References

14. L. Smolin (2007), *The Trouble with Physics: The Rise of String Theory, the Fall of Science and What Comes Next,* Houghton Mifflin Co., Boston.

157. H. Putman (1971), *Philosophy of Logic,* New York: Harper and Row, London: George Allen and Unwin, 1972. ISBN 0-04-160009-6.

158. M. Frede (1975), *Stoic vs. Peripatetic Syllogistic,* Archive for the History of Philosophy 56, pp. 99–124.

159. J. L. Bell (N/A), *Lectures on the Foundations of Mathematics,* THE PHILOSOPHY OF MATHEMATICS.

160. A. D. Irvine (2003), *Principia Mathematica (Stanford Encyclopedia of Philosophy),* Metaphysics Research Lab, CSLI, Stanford University. Retrieved May 8, 2009.

161. K. Devlin (1993), *The Joy of Sets (2nd ed.),* Springer Verlag. ISBN 0-387-94094-4.

162. J. Ferreirós (2007) (1999), *Labyrinth of Thought: A history of set theory and its role in modern mathematics,* Basel, Birkhäuser. ISBN 978-3-7643-8349-7.

164. T. Jech (2011), *Set Theory,* The Stanford Encyclopedia of Philosophy (Winter 2011 Edition), Edward N. Zalta (ed.), URL = http://plato.stanford.edu/archives/win2011/entries/set-theory/.

165. P. Koellner (2011), *The Continuum Hypothesis,* The Stanford Encyclopedia of Philosophy (Spring 2019 Edition), Edward N. Zalta (ed.), URL= <https://plato.stanford.edu/archives/spr2019/entries/continuum-hypothesis/>.

167. P. Suppes (1972), *Axiomatic Set Theory*, Dover Publications Inc.

174. J. R. Movellan (1995), *Tutorial on Axiomatic Set Theory*.

175. J. Nolt, D. Rohatyn, A. Varzi (1998), *The Theory and Problems of Logic (2nd ed,)*, Schaum's Outline Series: McGraw-Hill.

176. V. Borschev and B. Partee (2004), *Mathematical Linguistics,* Logic, Section 1: Statements of Logic.

177. M. M. Dougherty (N/A), *Mathematical Logic and Sets,* Department of Mathematics Southwestern Oklahoma State University.

178. P. Suppes (1957), *Introduction to Logic,* D Van Nostrand Co.

179. S. Lipschutz (1964), *The theory and Problems of Set Theory and Related Topics*, Schaum's Outlines Series: McGraw-Hill.

250. A. Miller (2016), *Realism*, The Stanford Encyclopedia of Philosophy; Edward N. Zalta (ed.),
URL = https://plato.stanford.edu/archives/win2016/entries/realism/.

251. Ø. Linnebo (2013), *Platonism in the Philosophy of Mathematics*, The Stanford Encyclopedia of Philosophy (Winter 2013 Edition), Edward N. Zalta (ed.),
URL = https://plato.stanford.edu/archives/win2013/entries/platonism-mathematics/.

252. G. Rodriguez-Pereyra (2016), *Nominalism in Metaphysics*, The Stanford Encyclopedia of Philosophy (Winter 2016 Edition), Edward N. Zalta (ed.),
URL
=https://plato.stanford.edu/archives/win2016/entries/nominalism-metaphysics/.

256. J. Bagaria (2017), *Set Theory*, The Stanford Encyclopedia of Philosophy, Edward N. Zalta (ed.),
URL = https://plato.stanford.edu/archives/sum2017/entries/set-theory/.

284. K. C. Klement (N/A), *Propositional Logic*, The Internet Encyclopedia of Philosophy.

285. Libretexts (2006), *Link between Logic and Set Theory,* Retrieved May 26, 2021,
URL = https://eng.libretexts.org/@go/page/10722.

286. Highbrow (2021), *Problem of Universals: Realism vs. Nominalism,* https://gohighbrow.com/problem-of-universals-realism-vs-nominalism/.

287. J. O. Wisdom (1953), *Berkeley's Criticism of the Infinitesimal,* The British Journal for the Philosophy of Science 4, 13, pp. 22–25. The University of Chicago Press.

288. W. C. Salmon (N/A), *SPACE, TIME, AND MOTION: A Philosophical Introduction,* University of Arizona, University of Minnesota Press, Minneapolis.

289. H. Robinson (2020), *Substance,* The Stanford Encyclopedia of Philosophy (Spring 2020 Edition), Edward N. Zalta (ed.),
URL =
https://plato.stanford.edu/archives/spr2020/entries/substance/.

290. S. M. Srivastava (2001), *The Completeness Theorem of Gödel:*
2. Henkin's Proof for First Order Logic, RESONANCE.

291. L. Horsten (2019), *Philosophy of Mathematics*, The Stanford Encyclopedia of Philosophy (Spring 2019 Edition), Edward N. Zalta (ed.),
URL   =
<https://plato.stanford.edu/archives/spr2019/entries/philosophy-mathematics/>.

292. J. Lambek (2017), *Foundations of Mathematics*, Encyclopedia Britannica,
https://www.britannica.com/science/foundations-of-mathematics.
Accessed 27 August 2021.

www.ingramcontent.com/pod-product-compliance
Lightning Source LLC
Chambersburg PA
CBHW060054100426
42742CB00014B/2832